S0-CAH-157

Recent Advances in Chemical Information

Recent Advances in Chemical Information

Edited by

H. Collier
Infonortics Ltd., Calne, Wiltshire

Proceedings of the Montreux 1991 International Chemical Information Conference, Annecy, France, 23-25 September 1991

Special Publication No. 100

ISBN 0-85186-496-1

A catalogue record for this book is available from the British Library

Published by The Royal Society of Chemistry,
Thomas Graham House, Science Park, Cambridge
CB4 4WF

Printed by Redwood Press Ltd., Melksham,
Wiltshire

Preface

The 1991 International Chemical Information Meeting and Exhibition took place this year in Annecy in France 23–25 September 1991. This was the third such annual meeting and, as before, some 200 attendees were present.

The 'Montreux' meetings began in 1989; their subject area has always been the latest developments in chemical information in electronic form. Chemical information — in a broad sense — has always had something of a pioneer role in the information world, due partly to its acute needs for highly specific searching and highly flexible retrieval, and partly to the fact that the large chemical, petroleum and pharmaceutical companies of the world have always placed a high value on good, accurate, high-quality information and have been prepared to support innovative new products and services in this area.

The 1991 conference featured a comparatively large number of interesting papers in the area of biotechnology, in addition to the 'traditional' papers in general chemical information and in patents. Biotechnology is an area that arouses increasing worldwide interest, and this year's conference faithfully reflected that fact.

The papers that make up the conference proceedings are assembled during the six week period before the conference; this gives authors the maximum possible time to write their papers, and also ensures that the information given is as up-to-date as possible. Those who have ever been concerned with publishing will know that, in the real world, there is a definite trade-off between maximum currency and maximum scholarship. The papers are therefore presented in this present volume in the same basic form as they were written by the authors during the summer of 1991. This means, alas, the occasional textual error and the absence of in-depth indexing. But it also means that the information is as current as time and technology permit.

In the early 1990s, the three major information areas that are significantly re-evaluating options in the light of modern technology are the areas of company information, the area of training and home information systems, and the area of chemical information. This current volume provides an interesting picture of just how modern information technology is beginning to affect the complex areas of chemical and patent information.

Harry Collier

Infonortics Ltd.

Calne, Wiltshire, England

Contents

The future of chemical information: information specialists versus end user searching

Clemens Jochum

Beilstein Institut, Frankfurt-am-Main, Germany

Abstract. *Currently the research information needs of bench chemists (i.e. end users) from electronic databases are mostly satisfied by company-internal information specialists. This is in contrast to printed information systems, which are generally searched by the end user directly. This paper tries to determine the reasons for this historical change in information behaviour. The outlook suggests that information specialists will always be necessary to carry out sophisticated search queries in complex databases. New easy-to-use PC- front-ends, databases on CD-ROM and different pricing structures of online databases will bring an increasing number of end users to satisfy most of their information needs without the help of intermediaries.*

1. The electronic age

Large information systems have been distributed as printed media for many decades — Beilstein for even more than a century. Although the demand for these printed systems is still strong, the demand for electronic access to the same information systems is constantly increasing.

In the late sixties and early seventies the American Chemical Abstracts Service was one of the first organisations to recognise the need for electronic access to chemical information systems. Others (such as Derwent and ISI) followed soon after, while Beilstein started its work on an electronic database only in 1982. Gmelin started even a year later. While Beilstein has now been available online since the end of 1988, Gmelin will probably go online by the end of this year.

Other information systems such as Landolt-Börnstein or Houben-Weyl are considering a computerised version of their information system but have not yet started.

2. The change in information behaviour

It is interesting to note that almost all of the electronic chemical information systems are publicly available in only one form — as online databases. These databases are searched mainly by only one professional group — the information specialists.

This is very different to the user behaviour of printed media. Traditionally the laboratory chemist himself conducted the searches in all printed information systems he needed. The librarian provided a necessary infrastructure (library room, ordering of information systems, help in locating these systems, etc.).

With the advent of the first online databases this search behaviour started to change. The librarian was no longer only a provider of information but became increasingly the intermediary who carried out electronic searches defined by the bench chemists. This not only implied a change in the librarians but brought about a whole new group of training professionals, the information specialists. The information specialist in chemical industry typically has an advanced degree in chemistry (very often an M.S. or a Ph.D.) and a thorough background in computer science.

Despite their important role as an information intermediary no special training is provided for them at most universities. In this regard the situation is worse at German universities than at American academic institutions but even there it is far from being satisfactory. Most universities have very good chemistry and very good computer science departments but their computer and electronic education for chemists is almost non-existent. The author presently teaches a one hour class on electronic information for chemists at the Technical University of Darmstadt. This class is not part of the chemistry students' standard curriculum and it is the only such course offered at this university.

Again, this situation compares unfavourably with past practice. For printed information products every chemist has to pass a quite thorough formal training in retrieving information from journals and secondary or tertiary literature. At some American and all British universities students had to take German language classes just to be able to read Beilstein! (This is no longer necessary since with its 5th Supplementary Series the Beilstein Handbook switched to the English language in 1984.)

The picture seems to emerge: chemists are doing less and less information searching themselves but leave this task to the information specialists. Universities will offer specific curricula for information specialists. Chemists, in turn, will soon be no longer educated in any kind of information retrieval.

Will this be the trend of the future? What are the reasons for this trend?

3. The need for information specialists

In 1984 the Beilstein-Institute conducted a worldwide study on the information needs and use in organic chemistry. 20,000 questionnaires were sent out worldwide and more than 4000 answers were received mainly from Western Europe, the USA and Japan. The results clearly show that information specialists do most of the database searching [1]:

- 53% of the respondents stated that in the organisations in which they work, there are centralised information services in which specialists carry out information retrieval for the research chemists.
- Information users from institutions having a centralised information service were asked to indicate what percentage of their information needs is supplied by these services. The results were:

Information needs	information users
up to 30%	56.6%
30 to 60%	25.2%
above 60%	18.2%

- The frequency of database searches by end users was very low. Only 16.4% indicated that they carry out searches regularly themselves (this already includes searches in internal databases).

As mentioned above, electronic databases demand totally new skills from their users. An experienced online searcher needs:

- Some basic computer knowledge: keyboard typing, knowledge about the hardware and software components of a computer, knowledge about programming languages is helpful for producing script files, etc.
- Knowledge about telecommunications
- Knowledge about database management systems to set up efficient search strategies
- Familiarity of the various search commands at the various online hosts (keep in mind that each host has a different command set. The command sets are about as similar as DOS and Unix commands)
- Database-specific knowledge. Familiarity with the database structure and the underlying chemistry.

To train all bench chemists of a company to acquire these new skills can be very expensive:

- The computer training courses are expensive. PCs have to be provided for all chemists.

- Search training is costly and time consuming. In contrast to printed material there is not a one-time charge for the material itself, but the charge goes up proportionally with the time needed for training.
- Bench chemists do not use the electronic search services on a regular basis. Because of the complexity of the search commands it is hard to maintain a high training level which is mandatory for efficient (and thus cost-efficient) online searching.

All the above reasons show that it seems to be more efficient to let information specialists do the searching.

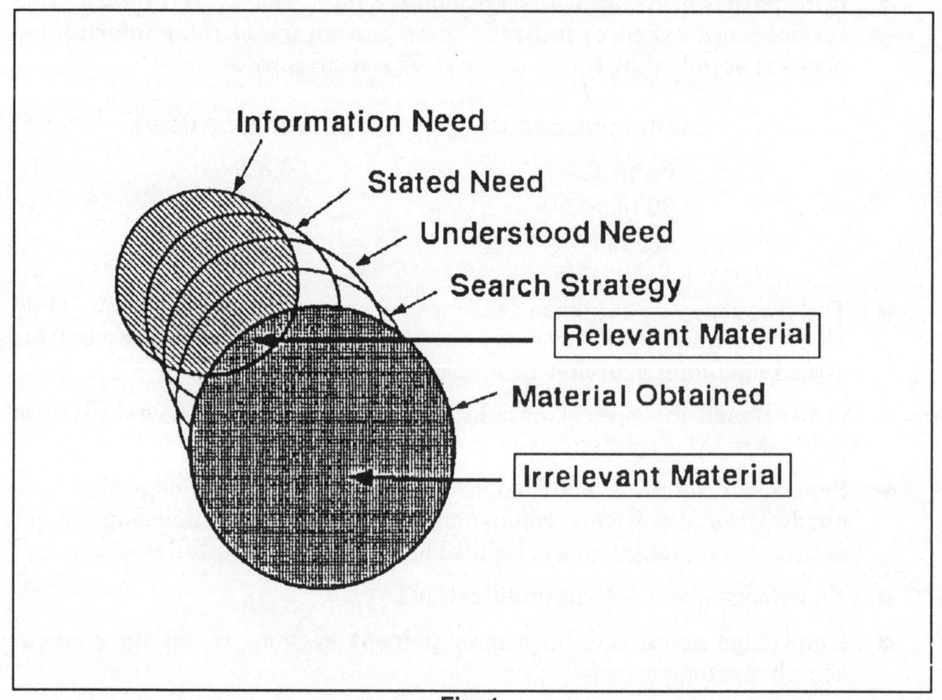

Fig. 1

The process of finding information through intermediaries. The end user needs a certain set of information ('Information Need'). By describing it to the information specialist, a small part is already lost. (Disjoint area between 'Information Need' and 'Stated Need'). Since every two people think slightly differently, the information specialist understands the need somewhat differently from what the end user means. Again information is lost. (Disjoint area between 'Stated Need' and 'Understood Need'.) The search strategy can in most cases only partially reflect the questions asked by a chemist who is not familiar with online searching. Again information is lost. (Disjoint area between 'Understood Need' and 'Search Strategy'.) In most cases the answers received from the search queries are not fully satisfactory. In summary, the overlap between the 'Material Obtained' set and the 'Information Need' set can be quite small, because information is usually lost at each step of the information retrieval process.

4. Reasons for End User Searching

Despite all the arguments mentioned above, a scenario where scientists (or other professionals) retrieve information through intermediaries *only* does not seem to be completely satisfactory:

- Approximately 25% of the total cost of newly developed products in chemical industry go into research and development. 40% of the inventions reported have been inadequately researched: 50% of these (i.e. 20% of the total) have been published before. This means that at least 20% of the total industry R&D budget is wasted every year due to inadequate information.
- More and more companies are offering database searching directly to their bench chemists. In-house chemical database management systems such as MACCS or REACCS are very successful and are used mostly by bench chemists directly.
- Whenever information is searched through an intermediary, information is usually lost. The relevant material consists of the intersection of the 'Information Need' Set with the 'Material Obtained' Set. The remainder of the 'Material Obtained' is irrelevant (Fig. 1). Equally important, serendipity becomes relegated and devalued as 'false drops'. You cannot delegate creativity!

The database with the worlds largest revenues by far (within all professional disciplines including chemistry) is almost exclusively used by end users and not information specialists: almost every lawyer in the US has a connection to *Lexis* (Mead Data Central).

5. Will the information behaviour change?

The last two sections show that on one hand the information specialists are currently handling most of the information needs of the bench chemists but that there are good reasons for end users to perform searches themselves. Are the end users going to put the information specialists out of a job?

Let us look again at the reasons why end users are not doing (in the main) their searches directly:

a) The cost of online searching is highly dependent on the kinds of questions asked, the search strategy and thus the skills of the searcher. The cost is therefore very difficult to predict and to keep within budget limits.

b) The translation of a non-trivial user question into an efficient search query is very difficult for two reasons:

b1) online search systems are not user friendly.

b2) databases have a very complex structure.

c) There is not much pressure from the end user side because the end users are not adequately informed about the potential of the databases or they do not consider the return high enough to balance the investment in time and effort required to learn the method.

Reason a) could be eliminated by changing the pricing of the databases. In the near future we will see more and more databases which will offer 'lump sum pricing', i.e. the customer pays a fixed charge per month or year for infinite searching (or up to a certain maximum).

Reason b1) could also be eliminated. We will probably see more and more user-friendly PC-front ends which will make it simpler for the occasional searcher to set up simpler searches himself. Programs like STN-Express (CAS), Molkick (Beilstein) or ChemMax (Maxwell Online) already are going in this direction.

Reason b2) will probably always require information specialists to handle either complex queries (such as patent claims) or to search complex databases. In this context a complex database is not necessarily a database with a complex data structure. In the author's opinion an end user can be trained more easily to search a factual database with a complex numeric data structure (i.e. with many numeric data fields) than to search efficiently a full text or bibliographic database. Soft data such as unformatted text tend to generate a lot more overhead if the query is not set up in an efficient and sophisticated way. With hard data and factual fields one has usually very little chance of ambiguity.

With the establishment of easy-to-use in-house database management systems for structures and reactions, more and more bench chemists are already satisfying a large part of their information needs themselves. In addition more and more databases are offered on CD-ROM for unlimited in-house use. Chemists can search them at their own pace without any cost pressure. These improvements will slowly eliminate reason c).

In conclusion it is out of the question that information specialists will always be an absolute prerequisite for complex queries and searching complex databases. Modern easy-to-use PC front ends, in-house database management systems, different pricing structures for online databases and more and more databases on CD-ROM will bring an increasing number of end users to search for the required information themselves. End users will not only search more databases directly but will probably use more and more information systems. This will lead to much less duplication of research results and this results in considerable savings on a company's R&D budget.

Reference

[1] C. Jochum, P. Moricz: *Database*, 41–46 (August 1987)

Training challenges of online chemical databases: Views of a producer, host and occasional user

Richard Kurt

ORBIT Search Service, McLean, Virginia, USA

The number of challenges confronting information producers and online hosts are legion, and none more so than in the area of chemical information. There are challenges too for the user, the person being subjected to the training of the producer or host. If these challenges can be properly met, the rewards for all parties can be very great indeed, ranging from a loyal and dedicated client base, to a new product for the producer or host to a cost and time efficient member of staff for an employer. These rewards are very real.

Before dealing with the challenges of chemical database training I would like to give a brief overview of the challenges involved in any training programme and some of the requirements that need to be fulfilled to ensure an effective one. There are two major components that will ensure a high degree of learning. The first is the learning process and how people learn. This is directly interrelated to the second component which is teaching skills or how to best facilitate the learning process.

How do we help students with the learning process? The first major objective is to set the students' expectations. Setting expectations has two functions. The first is to give them a goal or objective, something to strive for. This can be done very simply by giving an overview of the course, how it will be conducted and what the student will know by the end of the class. By setting these expectations you can motivate the student and tell them the benefits they will gain. In the online context you can set expectations with phrases such as: "You will be able to do conduct a complete structure search" or "You will use this to find reaction information to solve the problems confronting you". The second function of setting expectations is to put the student at ease and convey how the class will be conducted. By removing fears and relaxing tensions you can create a better atmosphere for the learning process.

Motivation, which we have just mentioned, is another factor that can help the learning process. A benefit statement can be a great motivational factor. Enthusiasm, challenges and praise can also be used to motivate. Finally we all know that acquiring knowledge is a rewarding and motivating experience. An instructor who uses motivation can increase and enhance learning.

Motivating students creates interest and interested students will pay better attention and thus absorb more. Interest and attention can be gained by other means. Your interest in them and your desire to be successful and for them to be successful can be critical. Boredom is a sin and students will miss important facts.

Another critical factor in the learning process is the media used. People learn what they know 75% through sight and only 13% through hearing. What does that say of a presentation without pictures? More dramatic is what a student retains, dependent on the media. Using words and pictures a student is likely to retain twice as much as using pictures alone and six and a half times as much as with words alone.

Analogies and logic also help the learning process. Going from the known to the unknown via the use of analogy relates new ideas to familiar facts, i.e., you are establishing associations. The use of logic can also lead from the familiar concept to the unfamiliar one and help create new paths. Logic teaches general principles. If a student knows the general principle he can continue to use it with a whole range of problems. It is easier to remember and apply one logical principle than a great number of individual rules.

In the United States, Missouri is known as the 'Show Me State'. Not until you are able to show a Missourian a fact, will they believe you. Immediate application of what has been taught is a method of showing. In the online context, showing is a hands-on session which allows the student to practice and test what they have learned. This increases credibility, and allows the student to participate which again increases the learning curve.

Immediate application is also a form of repetition, the last facet of the learning process, which will be dealt with. Considering that people retain only 25% of what they learnt after one week, repetition can be critical to ensure that the material sticks. Note that repetition does not necessarily mean repeating the same thing during the course of an instruction class but should include the chance for the student to review (and repeat) what he has learnt after it. In online training this is usually done by giving free practice time after a session is over.

The flip side to providing enhanced learning is effective teaching. There are many, many techniques to good teaching, some of which have already been mentioned, such as using logic, repetition, enthusiasm and setting goals. There are also numerous bad habits in teaching which can effectively ruin a whole session and I am sure we have all been on the receiving side of some of these sessions. I have also been guilty of perpetrating all of them and then some.

I would like to concentrate on five of the major skills: Knowing your audience, Knowing your message, Presentation style, Audio/Visual and Listening.

Knowing your audience is critical. If you don't, how can you effectively communicate with them? How do you ensure that boredom will not set in? How will you establish contact with the group as a whole and each individual. Without knowing your students you cannot even be positive that you will be covering all the pertinent material. Knowing your audience, starts with the class description, syllabus and publicity. A clear and accurate description 'pre-qualifies' your students. As an instructor, you do not want to have to cope with people at varying levels of knowledge. You will bore some and lose others. The fact that they are coming to the class and have read the syllabus or publicity material means that there is something there of interest to them. Further qualification of the students can be done at the class during the introductions — get the students to introduce themselves, describe what they do, what they know and what their expectations for the class are. All this information will help you to emphasize points or neglect others enhancing the learning process.

Knowing your message is the second critical factor to teaching skills. Your message contains the important bits of information you need to convey. Your message must be organized, clear, coherent and uncluttered. Organized means that you do not teach running before walking. Coherent and uncluttered means being focussed — teaching what needs to be known as opposed to what is nice to know. Being unprepared or unknowledgeable is a deadly sin in teaching since you will immediately lose the students' respect and ultimately their attention. Knowing your material will also enable you to set the pace of the class, allow you to digress dependent on the needs of the class and deal with questions knowledgeably and at the appropriate time. Sometimes you may wish to deal with a question at a later point. A tip when you are going to put off a question to later is to write it up in full view of the class as a reminder to one and all that it will be dealt with. Finally, you cannot know everything! If you have a question to which you do not know the answer, say so! Giving incorrect information is not the point.

Presentation style incorporates many skills all of which can enhance learning. Many of these are known to you, but let me mention a few: Your appearance can enhance or detract in teaching; eye contact can establish rapport and keep attention; voice projection and variety can make you easier to listen to; gesture can be used to emphasize important points and aid retention; sincerity will help with credibility; humor will make it fun and again assist in learning; vocal rate will allow your message to be absorbed. Pausing is an oft neglected skill but can be used to very great effect.

We have already talked about retention rates of words versus pictures versus words plus pictures. Effective use of *Audio/Visual equipment* is another important teaching skill. There are many kinds of audio/visual aids from a chalkboard to videos and datashows, each with its own advantages and disadvantages and each requiring different techniques for effective use. A proper discussion of these alone would require a full day but what we need to remember is that proper use of visual aids with words increases retention to 87% versus 7% for words alone.

Listening is the final but often neglected skill that I would like to touch on. If you do not listen to a student's query or concern you will not be able to answer effectively. Listening means full comprehension and trying to understand all aspects of the question. This may mean echoing back the query but using your own words to ensure your full comprehension of the problem. A question can indicate a learning block. You may have to address the topic from a different direction; never repeat word for word an answer. Listening and skillful answers should enhance respect and be an exchange of ideas. Good listening can also teach the instructor better methods of presenting materials or indicate a problem area with a product, which can in turn suggest a product improvement or a new product or service. There has been nothing more rewarding for me, in my role as an instructor, than to get a lot of questions since it means that people are paying attention and want to learn. They are also conveying important information to me through their questions. A class without questions leaves me wondering what I did wrong.

I now would like to relate this to online chemical database training. There is a challenge -- in the words 'online' and 'chemical'. Does a producer of chemical information need to teach online? Does a host need to teach chemical input? How should the course be structured to benefit the students? And what about the students? How much chemistry or database or online knowledge do they have? Are they end users? What needs to be conveyed to make them effective users? Answers to these questions can be found if we relate them back to the learning process and teaching skills. Let me provide a case history, which in many ways shows the evolution process in online chemical databases and training. It will also show the rewards that can be gained.

For a number of years I was involved with an organization that had a sophisticated indexing system for chemistry and polymers. Note that sophisticated can mean complicated! Training was conducted over several days. This was already a bad sign. Was the system that complicated or the teaching that poor? A further complication was that there were virtually no specific materials for the class apart from the indexer's manual, a sheet of questions and answers and a questionnaire. Needless to say the classes were very difficult for both instructor and student alike and the completed questionnaires left very little doubt that improvements were needed. The net result of the classes was students who not only had the greatest difficulty in using the system but also harbored vengeful feelings for being subjected to the training.

Two basic problems to the learning process were identified. The first was that users were having an inordinate amount of problems in time ranging. The code had been modified and enhanced a number of times over the years. To make the best use of the code and to reduce false drops, these modifications had to be incorporated in the search logic. This involved a multi-step process and there was more than one way to ultimately get the correct solution. This added to the confusion since we tried to cater for all these variations rather than giving one simple, logical format. Finally, selection or placement of a code in the wrong step could have a drastic effect, limiting the results from a specific period onwards. It was very easy to lose five years worth of data. Yet the whole system had been devised for complete and comprehensive retrieval.

The second problem was the indexer's manual. A user coming to this had immense problems with the organization and finding the correct terms. Cryptographic notations for the indexer were unhelpful in assisting the user to construct logics. I can laugh at it now, but a half hour was spent in teaching a conversion from a year to a letter and then back again since these letters were used throughout the manual to indicate which year a code was introduced. Ensuring that all relevant codes were found to a search question was another problem.

Solutions to both these problems were found and both were intrinsic not only to better training but also resulted eventually in new products which not only assisted the user in learning but also in using the code — the ultimate goal of our training.

The first part of the solution was a search strategy sheet which guided users line by line through the year ranging problem. Each line corresponded to codes from a certain period. This greatly facilitated the teaching process, making a topic that could take a day, explainable in a half hour. The second part of the solution was a user oriented manual — a dictionary. The dictionary was not only an alphabetic listing of concepts with associated codes and replete with cross-references but was related back to the strategy form. Codes were numbered to indicate placement on the form. Perhaps the best way of looking at the two components is that we had created a non-computer based menu system. The whole system had been structured and organized for the user.

The package reduced training from two days to one, but its biggest success was that users went away knowing how to use the code effectively and if not loving us, at least feeling a sense of accomplishment. Note that while we were enhancing the learnability we were also sharpening our teaching skills.

The story is not yet at its end. The PC was becoming a significant tool in our industry. The dictionary and strategy form were loaded as a true menu software. The program was unfortunately rather tedious to use, with screen after screen of information that had to be scanned but the company was well into its next upgrade — a system that translated a chemical structure to codes and constructed the logic, time ranging and all. The new product was a winner on all sides. It simplified teaching, made the learning easier, made a complex area more accessible to users and finally is used as a sales tool to bind and win customers new and old.

Have I exaggerated the importance of training in the development of this product? Well maybe slightly, although I can take no personal credit. Have I exaggerated the importance or challenges of training otherwise? Not at all, I believe. In fact, I believe I have seen analogous developments by other organizations. Training should be a very high profile area for any organization in the online arena since the rewards can be so great. Will training open up the vast gold mine of end users? Not yet is the answer. Numerous studies have indicated that while end user training is beneficial, there has been no great leap in usage. Reasons range from fluctuating information needs, i.e., they user does not use the system enough to stay current, to equipment not being available, to there being little encouragement given by management, to more training is needed. To me, this last point indicates that online systems and chemical databases are still complicated to be used by the non-specialist.

Training will however continue to play a significant role for online chemical databases. It will also influence new products and systems and win or lose customers. Not until a user can go online and ask an unstructured question will need for effective training disappear. That will not happen for a long time.

New access modes to reliable computerized numeric properties data

V. J. Drago and J. G. Kaufman

Chemical Abstracts Service and MPD Network, 2540 Olentangy River Road, P. O. Box 3012, Columbus, Ohio 43210, USA

Abstract

The increased availability of quantitative properties data for chemical substances and compounds and of special tools for accessing, searching, and retrieving such data has raised the level of activity in this area. In this paper we will discuss the nature of numeric data and how they differ from bibliographic data and preview newly available and forthcoming tools that make locating and retrieving such data more attractive.

Numeric data differ from textual data both in their nature and in the means of their representation. Numeric chemical information can be for single substances, for static mixtures, or for reactions. They have associated variables upon which the values of the properties may depend and associated units that must be detailed along with the values themselves.

A variety of properties databases are now available on STN International, the scientific and technical information network, including Beilstein, DIPPR, JANAF, NISTTHERMO, and NISTFLUIDS. In addition, a number of enhancements have been incorporated into the MESSENGER host software specially tailored to searching numeric data, such as range searching, tolerance setting, units conversion over six systems, and table display. Additional assistance to the searcher for properties data has been provided by the addition of NUMERIGUIDE, a file the user may consult to learn which files are most likely to contain the type of data sought, thereby eliminating time wasted looking in files without the desired property.

Perhaps of most interest to the engineer or scientist looking for the property data is the forthcoming addition of a menu-driven interface to this cluster of numeric files. This interface will enable the searchers to search all of the files with very simple, self-guiding menus and either to select their own search files or let the interface make the file selection based upon the query.

Introduction

A large number of excellent online bibliographic or textual files have been available for some time, but the development of properties files through which engineers and scientists can gain direct access to performance properties of chemicals and structural materials has come only in the past decade. A number of expert studies in the early 1980s [1,2,3] focused attention on the critical need for easier access to such information, and it has taken the rest of the decade to begin to implement major system responses to those needs [4,5].

The main reason for the slower development of numeric properties databases is the relatively complex nature of such databases. In this paper, we will examine the nature of numeric data as contrasted to textual or bibliographic data. Because of the distinct nature of numeric data, specialized search and retrieval software has been developed to enable searchers to locate specific answers to rather sophisticated questions, and our review will include some of those technologies.

The user base itself has also influenced greatly the way numeric data are dealt with in commercial databases as the needs of the so-called end-user — the individual who will actually use the retrieved answers — are taken into account [6]. Unlike the information professional, who spends all day searching and can afford to learn several complex command languages, the end-user engineers or scientists can afford to spend only a small portion of their total effort searching for such results. The usual alternative of passing the request on to an information professional is often unattractive, unless that individual has considerable scientific depth and experience, because the queries are so specific and sophisticated. As a result, special menu interfaces are now more often available. Though tedious to the information professional, they permit the end-user to search with little or no prior knowledge of the database management system and to search an entire group of databases, not just one or two. It is useful to look at progress in the development of such end-user interfaces as part of this study as well.

Nature of numeric property data

Numeric property data are notably different from textual data in several basic respects. First the obvious: they are quantitative and thus have implied precision, come in ranges as well as discreet values, and may vary over many orders of magnitude within a single data record containing four or five properties. Second, they have units associated with them from which they can never be separated and still retain meaning, and the numbers in any one record may have multiple units, some of them extremely complex. Finally, they are usually dependent upon a variety of variables or combinations of variables, usually called parameters, which influence their value or behavior in certain situations.

These factors all complicate the handling of numeric data beyond that for textual data, which typically is searched for 'strings' of characters without other modifiers except perhaps other strings that may be added together in various boolean modes. Numeric data must be stored and searched keeping in mind at least three major elements — name of the property, its value, and the units of that value — and often more. For example, the property may be dependent upon time of exposure at a particular combination of temperature and pressure; therefore, three more factors — time, temperature, and pressure — must be included as delimiters in every query about that property.

Numeric data software capabilities

In order to provide search and retrieval software responsive to these needs, operators of scientific online systems like STN International have over the years built into their base software certain capabilities that are targeted at handling numeric data:

- Range searching — the ability to search for materials with combinations of properties in specific ranges or above or below certain limiting value;
- Units conversion — the ability to convert to any of the worldwide standard systems of units: International Standard (SI), meter-kilogram-second (MKS), centimeter-gram-second (CGS), engineering (ENG), and the STN user-friendly SI system;
- Tolerance setting — the ability to define ranges of search values by the tolerance on the search value (e.g., 50,000 plus/minus 1000 psi);
- Table display — the ability to obtain tabular display of data that match a specific query and to pre-define certain types of tabular displays;
- Calculation packages — the ability to interpolate and estimate additional information or to apply parametric analysis of multivariant properties, which is valuable in providing specific answers to some complex materials questions.

Enhanced interface

As noted earlier, the individual most likely to frame a very detailed query for specific performance data is often the end-user who does not recognize the cumbersome command languages of most online systems. STN International has already implemented one development interface for the materials files of the MPD Network and is currently working on several others. Among the major features of these interfaces are

- Logical, easy-to-use, menu-driven search paths;
- A variety of search paths, recognizing that different types of users and different applications will require different queries;
- A 'metadata' system in the form of an interactive thesaurus that both deals with user queries, translating them to all other acceptable nomen-

clature and terminology, and responds quickly to clarify the meaning of names, terms, and abbreviations;

- A directory of data sources, including those outside the MPD Network.

Depending upon the nature of a query, users may approach the file cluster with different pieces of information at the heart of their query:

- A specific database with a certain type of data, e.g., design values;
- A specific material for which a variety of types of data are sought;
- A specific property or properties for which a comparison of materials is required, perhaps involving a specific range of values, notably those equaling or exceeding certain limiting values.

Presentation formats

Similarly, providing for display of numeric properties data introduces another level of complexity. For each file, the following factors are considered:

- Types of users — engineers and scientists tend to expect data to be presented in a logical, use-oriented tabular or graphic display, rather than in a normal linear order of retrieval. Wherever appropriate, property data are presented in tabular matrix displays.
- Types of queries — special display formats, known as 'query-related' displays, are utilized to highlight specific properties and the variables or parameters that define the applicability of the data. Searchers do not have to wade through extraneous data, yet they have all of the pertinent related information.
- Number of answer sets — brief summary displays have been designed to allow searchers to scan large numbers of records to locate those of particular interest, which can then be displayed more fully.
- Documentation required to define the source and applicability of data — as noted earlier, a number of variables such as time and temperature or, in the case of the materials, form and dimensions are often required to define the range of utility of specific data. Displays are designed to accommodate reporting of these variables along with the specific data requested.

The provision of supporting documentation in the displays is a particularly important point often taken for granted or omitted completely in reporting data. This may go well beyond what a specific user requests. Basic facts such as the type of data (design value or individual test result) and the applicable orientation (longitudinal or short transverse) must be included.

MPD Network on STN International — initial application of cluster searching

The National Materials Property Data Network, Inc., was formed in 1985 with the mission of providing easier access for engineers and material scientists to reliable materials properties data. In 1987, the NMPDN, or MPD Network as it is known, joined forces with Chemical Abstracts Service (CAS) to implement and distribute the MPD Network service on STN International. The prototype production version of MPD Network became available early in 1991. The individual materials property files continue to be available to STN International searchers who wish to use the standard command language rather than the MPD Network interface.

STN International is the premier online scientific and technical information network. It is operated jointly by CAS, a division of the American Chemical Society in Columbus, Ohio, FIZ Karlsruhe, a scientific and educational organization in Karlsruhe, Federal Republic of Germany, and the Japan Information Center of Science and Technology (JICST) in Tokyo, Japan [7]. In 1987, MPD Network and the American Chemical Society initiated a joint development program that has led to the availability of MPD Network services on STN International.

A broad range of numeric information is already available on the MPD Network, and much more is scheduled in coming years as further described below. The intended scope includes not only mechanical, physical, and other performance data for all structural materials, including metals, polymers, ceramics and composites, but also the properties of connections and joints in these materials.

Among the specific databases available through the MPD Network are

- AAASD — from the Aluminum Association — typical and minimum tensile properties, typical mechanical and physical properties, and fabricating and application information on more than 100 commercial aluminum alloys
- ALFRAC — from NIST/SRD, the Materials Properties Council, and the Aluminum Association — plane strain fracture toughness data for about 25 high strength aluminum alloys, with validity documentation
- IPS (International Plastics Selector) — from D.A.T.A. Business Communications, a Division of IHS
- MIL-HDBK-5 — from the MIL-HDBK-5 Coordination Committee publication — design tables covering the design mechanical and physical properties of ferrous and non-ferrous alloys
- MARTUF — from Materials Property Council — about 10,000 individual toughness test results for steels for marine applications

- METALS DATAFILE — from Materials Information (ASM International and the Institute of Metals) — data from more than 40,000 literature citations from technical journals
- PLASPEC — from the publishers of Plastics Technology — typical properties from producers of about 10,000 plastics
- STEELTUF — from the Electric Power Research Institute and the Materials Properties Council — the results of more than 20,000 individual tests of steels for the power and petroleum industries

Programs are in place to continue adding about four databases to the MPD Network each year. Among the subject areas currently in development are

- Properties of ceramics and ceramic components
- Properties of copper and copper-base alloys
- Optical properties of glasses
- Properties of reinforced polymer composites
- J-integral values for high-temperature structural steels
- Creep and stress-rupture properties of metals
- Fatigue properties of metals and alloys
- Properties of lumber and other wood products

Specific database names will be announced as distribution arrangements are completed.

The Chemical Property Data Network

The next application of the cluster search concepts will be on the chemical and thermodynamic properties of chemical substances. Among the databases currently on STN International with direct application in this area are

- BEILSTEIN — a major structure and factual database of critically reviewed information in organic chemistry representing about 3.5 million cyclic, heterocyclic, and acyclic substances;
- DIPPR — textual and numeric information on the pure component physical property data for commercially important chemical substances;
- HODOC — an electronic version of much of the CRC Handbook of Data on Organic Compounds, including the chemical and physical properties of over 25,000 organic substances;
- JANAF — critically evaluated thermodynamic properties from the Joint Army-Navy-Air Force Thermodynamic Tables;

- NISTTHERMO (formerly NBSTHERMO) — the NIST Tables of Chemical Thermodynamic Properties, containing critically evaluated chemical thermodynamic properties of over 8000 inorganic and organic substances;
- NISTFLUIDS (formerly NBSFLUIDS) — the NIST calculation package covering the thermophysical and transport properties of fluids as a function of temperature and pressure;
- TRCTHERMO — from the Thermodynamic Research Center — the evaluated thermodynamic properties of 7000 compounds.

This cluster of files will soon be supported by an interactive menu-driven interface that, like the MPD Network interface, will aid engineers and scientists as well as professional searchers to handle the complexities of numeric data.

Close cooperation with international standardization activities

In developing the worldwide chemical and material property database networks, MPD Network and CAS not only work closely and cooperatively with many technical societies, government agencies (notably the Standard Reference Data Group at the National Institute for Standards and Technology), and private industry but also with many international organizations concerned with standards and quality and reliability issues. These include

- ASTM Committee E-49 on the Computerization of Chemical and Material Property Data in developing standard formats for recording, storing, and exchanging chemical and materials data [8]
- CODATA, the International Committee on Scientific and Technical Data of the International Council of Scientific Unions through its Task Group on Materials Database Management [9]
- VAMAS, the Versailles Advanced Materials and Standards group through its fact-finding studies and developmental workshops [10]
- ICSTI, the International Council for Scientific and Technical Information in its development of referral databases.

MPD Network and CAS are also in regular communication with other major scientific information projects such as the European Community Demonstrator Programme [11].

Summary

STN International, with the MPD Network and the newly forming Chemical Property Data Cluster, has taken an innovative approach to providing engineers and material scientists easy access to worldwide sources of numeric chemical and material property data. The distinct added complexities of storage, search, and retrieval of numeric property data have been addressed with special search software and a user interface to aid the scientists and engineers who are the end-users of such highly technical information. Provision has been made to ensure that a variety of types of users and their queries may be handled with ease on the system and that the confusion of names, terms, and abbreviations can be removed.

References

[1]. Ambler, E., 'Engineering Property Data — A National Priority,' *Standardization News,* pp. 46-50, ASTM, Philadelphia, PA., August, 1985.

[2]. Materials Data Management — Approaches to a Critical National Need, National Materials Advisory Board (NMAB) Report No. 405, September, National Research Council, National Academy Press, Washington, September, 1983.

[3]. Computerized Materials Data Systems, Proceedings of the Fairfield Glade Conference, J. H. Westbrook and J. R. Rumble, Editors, National Bureau of Standards, Gaithersburg, MD, 1983.

[4]. Kaufman, J. G., 'The National Materials Property Data Network: A Cooperative National Approach to Reliable Performance Data,' Proceedings of the First International Symposium on Computerization and Networking of Materials Databases, ASTM STP 1017, 1989, Philadelphia, PA., pp. 55-62.

[5]. Rumble, J. R., Jr., and F. J. Smith, Adam Hilger, Database Systems in Science and Engineering, Bristol, 1990.

[6]. Kaufman, J. G., 'Increasing Data System Responsiveness to End-User Expectations,' Computerization and Networking of Materials Databases: Second Volume, ASTM STP 1106, Philadelphia, PA., 1991, pp. 103-112.

[7]. 'STN International — Databases in Science and Technology,' American Chemical Society, Columbus, OH, 1990.

[8]. Rumble, John, 'Standards for Materials Databases: ASTM Committee E49,' Computerization and Networking of Materials Databases: Second Volume, ASTM, STP 1106, Philadelphia, PA., 1991, pp. 103-112.

[9]. CODATA Bulletin No. 69, 'Guidelines For Materials Database Management,' November, 1988, Hemisphere Press.

[10]. 'Factual Materials Databanks — The Need for Standards,' Report of VAMAS Technical Working Area 10, H. Kroeckel, K. Reynard, and J. Rumble, Editors, June, 1987.

[11]. Kroeckel, H. and G. Steven, 'EC Action Towards European Pilot Network for Materials Data Information,' Materials Property Data — Applications and Access, MPD-Vol. 1, PVP-Vol. 111, ASME, New York, 1986, pp. 175-182.

If only . . . A sideways look at some patent databases

J.F.Sibley

Shell International Petroleum Co. Ltd.

I am going to look at a few of the newer developments in the area of patent information from the point of view of a searcher working directly for one or more patent attorneys. I have worked as such for more than twenty five years and, in that time, there have been many changes not only in the methods used to retrieve data but also in the sources of information themselves.

When I started most large companies relied on in-house patent databases developed either alone or in conjunction with others as in the Patent Documentation Group. The major third party sources were almost always provided by the patent offices themselves although secondary services such as Chemical Abstracts and Derwent Publications Ltd. had their place. There were no online services; everything was in the form of hard copy although some Derwent material was available in the form of 80 column punched cards which could be searched using a single column card sorter. These sorters were invariably situated in the accounts department and, in my case, the cards in boxes had to be wheeled on a porters trolley along a country road and into another building. The sorter had a well developed appetite for Derwent cards and the number of cards brought back to the office was always less than the number taken out.

It has all changed now! Most companies have stopped producing their own patent databases and rely to a very great extent on those provided by third parties. We in the chemical area are very fortunate, not only are we awash with online and hard copy services, with microform and CD-ROM, but we are promised still more. But have matters improved? Well, in general the answer must be yes, many things are easier and much quicker to do and results can be more reliable. This is particularly true for bibliographic searches. For example, tracing patent families which used to involve working through the official name indexes from each country of interest together with a look at the Chemical Abstracts Patent Concordance, now requires only the online use of INPADOC together with Derwent's World Patents Index (WPI) which will give very acceptable results. The problems come with subject matter searches. Many in-house patent databases covered a limited number of technical areas and the subject matter could be indexed by a personal system accurately tailored to the needs of the company. Once this intellectual effort is stopped, reliance must be placed on the classification systems applied by third parties, amongst them the patent offices, and on the accuracy and consistency of their coders.

The existing patent databases are good, but none is perfect. All could be improved. It is extremely rare to get a search which can be safely carried out using just one database. At least two, more often three or more databases will have to be consulted to ensure satisfactory search results even for novelty and/or patentability searches. However, every possible source must be used for infringement searches, which are searches made to establish that a company is able to manufacture and market a product without fear of being stopped by the existence of one or more valid third party patent rights. Even then the search can seldom be exhaustive and there are usually lingering doubts about the completeness of the answer.

So any new patent database must be good . . . musn't it?

ESPACE products

The European Patent Office (EPO) has been very active in developing and marketing a range of CD-ROM products based on patent publications. The oldest of these is ESPACE EP-A which comprises facsimile images of the full specification of all EPO A-documents published since January 1989 together with the bibliographic data in character coded format. Other ESPACE products include the full texts of PCT applications, the front pages of all EP-A documents and also a bulletin product. Future developments will cover UK, German and Spanish patent documents.

At present, the CD-ROM products are being used mainly for archival storage, replacing collections of hard copy specifications. They could be used to run SDI in-house, but at present there are limited search possibilities although the International Patent Classification (IPC) symbols are available. If more front page data elements become searchable, such as words from the abstracts and especially the drawings and formulae, their usefulness for SDI will increase. This will be particularly true in the chemical area where a combination of the IPC with a structure and/or word search could give fairly specific retrieval. With the ESPACE EP-B CD-ROM, words from the first claim of the granted patent are already searchable, but these documents are of restricted value for SDI.

If the use of CD-ROM for in-house SDI does increase, it will pose a serious threat to those commercial organisations providing such alerting services at present, for the ESPACE products are very cheap and are published concurrently with, or very soon after the printed specifications so the commercial services will have to add considerable value to the raw patent data if they are to retain their market share. They do this at present by including formulae and drawings together with abstracts or significant key words.

I believe that the use of the ESPACE CD-ROM for retrospective searching is less likely to develop for not only are there hardware problems with regard to storage and access, but also the retrieval possibilities are limited. The title and abstract will be that provided by the inventor or his attorney and it is unlikely that they will be as informative as the corresponding Derwent title and online abstract. The IPC will be that provided by one examining office only whereas the corresponding Derwent record will carry the IPCs provided by all of the offices publishing the invention which, together with other Derwent coding, will usually give greater scope for retrieval.

As larger juke boxes or other storage means for such data are developed, together with cumulative indexes, the situation could change, but this appears to be some way off and I really cannot see the ESPACE CD-ROMs as anything other than an archive from which would be printed specifications found as a result of searches in online databases.

One CD-ROM product which could have considerable influence on European searchers is the CASSIS disc containing the bibliographic information files of US patents. This disc will provide a very useful and convenient entry into the US patent classification which is not the most user friendly system available.

Full text databases

We used to hear a lot about full text patent databases and how they would replace all of the existing abstract services, but they fell from grace when the problems of automatic translation of patent specifications became apparent. The only full text patent database of which I am aware is Lexpat which covers US patents. The World Intellectual Property Organisation (WIPO) intends to change all that.

In 1989, the WIPO Permanent Committee on Industrial Property, convinced that the most important matter for international co-operation in the field of patent information, at least in the next decade concerned:

(i) the storage of full texts, including drawings, of at least all new patent documents on optical discs or other devices capable of storing such texts in an extremely compact and easily accessible form;

(ii) the further development of highly automated and computerized retrieval systems in which retrieval is based not only on classification but also on words, combination of words, chemical formulae and other elements;

decided that the task of storage should be shared between certain offices using compatible hardware and software. Additionally, the search methods of all offices and commercial databases should be so harmonised that each of them could search the databases of the others. The highest priority is to be given to promoting the adoption by patent offices, other public institutions and private enterprises of such storage means for full texts (including drawings) of patent documents and of compatible search software [1].

This full text database, like the CD-ROM products, is a spin-off from the move by the EPO, the United States Patent and Trade Mark Office (USPTO), and the Japanese Patent Office (JPO) to paperless offices; the latter being some way down this road [2].

Additionally, the EPO, together with the USPTO and JPO are already capturing in facsimile mode the documents which comprise the PCT minimum documentation and this is to be extended later to include the documents of other EPC countries. More recent documents have been captured in mixed mode and it is thought that this mixed mode, multi-language database will be made available to the public by commercial organisations.

I have serious doubts about the commercial value of such full text storage systems, and I cannot see even the most perfect full text database providing a very valuable day-to-day search tool for anyone including the patent offices. I believe that the larger offices will continue to produce their own search files to carry out the necessary patentability searches. I am not alone, Stu Kabak, one of the most experienced searchers around, has expressed similar doubts [3].

As with the CD-ROM, with which the full text database will compete, the possibilities for successful use of full text searching will be highest in the chemical area where a combination of reasonably precise classifying codes with words from the text, could provide enhanced retrieval. But, even here there could be problems for, not only are inventors and attorneys allowed to use their own language when writing the specification, but also they are given a free choice of the words they use to describe every feature of their invention. Again, Stu Kabak has pointed out some of the difficulties that can arise with the naming of chemicals [3]. Generally, raw patent specifications do not provide a good searching medium, it is far better to use a database which has been deeply classified and indexed by knowledgeable people.

Of course, intelligent, accurate, automatic translation of free text may be introduced within the decade but it seems unlikely. One has only to look at the names of non-Japanese companies which have been translated firstly into Japanese and then back again into something approximating to English to realise how far off is automatic accurate translation. If sensible searching of full text patent databases is to be achieved then attorneys will have to be encouraged to use controlled language, and an online multilingual, hierarchical thesaurus will be required which will take any multinational committee at least a decade to produce.

In any case, bearing in mind the convergence in patent procedures and the increasing acceptance of a single language filing text, would you want all patent documents in a full text database or just one from each patent family? All-in-all, it may be better for the offices to co-operate with Derwent to get the complete Derwent Documentation Abstract online rather than spending time and money putting up full text databases which will be of limited value.

EPOQUE

EPOQUE is the new, in-house online retrieval system of the EPO. The EPOQUE command language will allow the EPO examiners to interrogate interactively not only the in-house databases of the EPO but also the databases of the USPTO and JPO and, additionally, Derwent's WPI database, which is also available in-house. Access to remote, third party hosts with automatic translation of the EPOQUE search language (based on that of Questel) into that required by the different hosts is also provided. This is all achieved by the EPOQUE workstation.

This is a most exciting and potentially useful development, but it is not going to be made available directly to the public. Rather, it is going to be provided, freely or at low cost, to the national offices within the EPC for them to use to disseminate patent information to their nationals. In the United Kingdom, as in some other countries within the EPC, the Patent Office is not seen as a major source of patent information; people turn instead to the Science Reference and Information Service (SRIS) based in the old Patent Office Library in London. Whilst the Patent Office does collaborate with SRIS to some extent in the dissemination of patent information in the UK, it seems unlikely that either party will be equipped or funded sufficiently to make the EPOQUE system widely available to UK entrepreneurs. Indeed, even if one of the EPOQUE workstations is placed in SRIS, it will be of little value to the bulk of British industry which is situated away from London. Certainly the ESPACE CD-ROMs which are available in SRIS have been very little used.

EPOQUE would be of far greater value to European enterprises if it were to be made available to them for their own use in addition to it being provided to the national offices and/or other responsible organisations. An annual fee could be paid to the EPO who could use the income to fund the development of patent information services in the less well developed countries of the EPC.

It is a fact that the majority of patent documents do not add significantly to the fund of human knowledge. The national offices could benefit their nationals most if they concentrated on selecting from the mass of patent documents those which appear to make an important contribution to a particular technology and then selectively bringing them to the attention of organisations working in that area. That such an approach is possible can be seen, for example, by looking at that part of the annual report of the Patent Office which focuses on current trends in patenting. Only by adding value in this way, will the national offices make a useful contribution to the informational needs of the EPC member states.

The EPO has recently taken over the Vienna-based INPADOC service and they have continued to make it available to subscribers but, according to the constitution of the EPO, should a national office within the EPC offer to provide the same service, the EPO will withdraw their service from the nationals of that country. This is something which causes concern amongst regular users of the high quality INPADOC service provided on the Vienna computer who would be reluctant to change to a service provided by their national office which may be dependant on unpredictable funding from central government.

Markush databases

There are three Markush databases available online but, as far as I know, they have not attracted very much attention in Europe although we are told that the Japanese and Americans are making greater use of them. WPIM, the structure searchable part of Derwent's WPI file, and MPHARM, the structure searchable part of INPI's Pharmsearch database both use Markush Darc software and are available on Questel whilst Marpat, based on Chemical Abstracts, uses software which is closely related to that already used to search the Registry file on STN. In addition, a fourth service is being developed by Sheffield University and the German International Documentation Company for Chemistry (IDC) based on the Gremas software developed by Sheffield University.

I have used both Marpat and M-Darc but do not use either on a regular basis. Various comparisons of the publicly available services have been made and reported [4-6], but I have no intention of going into these matters today.

It seems to me that when deciding to develop a Markush database, the producers must have a very clear understanding of the potential customers. From my discussions with colleagues in other companies, it appears that they do not anticipate that bench chemists will be given access to a Markush file; it will be too expensive and very few research scientists will appreciate the scientific and legal significance of any answers found. Similarly, academics should not be concerned with using such a database although they could usefully make far greater efforts to educate their students in patent matters. The potential number of regular users is very restricted, comprising mainly the patent offices and those searchers who work directly for patent attorneys whether they be in companies or private practice.

The market within patent offices may not be very large. The three major examining offices, the USPTO, the EPO and the JPO are moving closer together and they will co-operate by exchanging search reports when they are processing equivalent applications. This means that the total number of searches made by these offices could decrease substantially and, even for the residue, Markush files may not be used.

John Brennan, at the EPO seminar on Search and Documentation Working Methods in March this year, stated in a pre-print of his paper that "While Markush Darc, or Marpat, may be of increasing utility as the size of their databases increase, no significant use has been made of these systems to date". Both during his presentation and in his workshops later, he made the point that if prior art which, whilst making a broad disclosure of the new invention in fact gave no clear teaching of that invention, was not thought to destroy patentability, then there would be little incentive for EPO searchers to make use of the more expensive Markush files in addition to their normal databases.

The USPTO, when faced with very broad generic disclosures has a tendency to telephone the applicants' attorney who makes a selection of preferred compounds which are then searched by conventional means. If these are found to be novel and inventive, then the rest are assumed to be novel as well. I have no knowledge of the working practices within the JPO, but, in any case, the potential market within the major offices would appear to be quite small.

Much of the work done by outside searchers is to establish the patentability of a company's inventions, and there can be little doubt that many are going to be influenced by the EPO's views on patentability. Due to the advanced search software, the Registry File on STN is capable of giving acceptable results for patentability searches, especially when combined with a search of the EPO internal search database (EDOC) available on Questel and of the US patents using CASSIS. The Markush files will only become more interesting when carrying out infringement searches.

If the overall market is so limited, it is very important that the database producers ensure that they provide search facilities which meet, as closely as possible, the needs of this small group of users.

The search software must reflect the language used not only in the patent documents themselves but also in the queries that the searcher will receive from the attorney. There must be, for example, the possibility for using variable points of attachment, for translation between generic and specific compounds and *vice versa*, for providing for optional substitution, for optional unsaturation, for groups which may or may not be present, and for groups which can exist separately or join together to form a cyclic moiety. It is desirable that the query language should include provisions for some query fragments to be matched precisely, whilst other fragments can be matched with either specific groups or the corresponding more generic groups. At present, there is no perfect search software although the Marpat query language already provides many of the features listed and proposed future developments for this system look very promising.

However, it is no good having perfect search software if the data in the database are incomplete or inaccurate. The present commercial Markush systems have limitations, for example, on the number of variables that can be present and on the number of alternatives for each variable that can be handled. In addition, some chemical subject areas are not included in the databases.

To be of value, all patent documents containing Markush formulae must be covered by the databases regardless of the technical area of the invention and regardless of whether the information is being published for the first time or not. All the disclosures made in the specification must be included, it is not enough just to use the claims as the basis of input.

Where the input and search software allows for generics such as 'alkyl', 'alkenyl','aryl' and 'heteroaryl', the definitions of these generics should be the same as those generally accepted by the patent courts. The input must reflect accurately the disclosures, there is no room for inference on the part of the encoder. In many patent documents, the Markush formula contains 'aryl' substituents but, in the body of the specification, 'aryl' is defined as including, for example, phenyl, naphthyl and pyridyl, so the database input must include both of the system generics 'aryl' and 'heteroaryl'. It is permissible to use only 'aryl' as input where there is either a clear teaching that only carbocyclic rings are envisaged or there is no definition of the grouping at all. The language of the chemist must take second place to the language of the attorney when dealing with patent documents. I have seen a specification in which the Markush formula contained 'alkyl' substituents, seemingly straightforward until you read the specification in which the attorney had defined 'alkyl' as embracing 'alkyl', 'alkenyl' and 'alkynyl' and the database record must show this. In the same vein, there are cases in which a substituent R has been defined as 'a saturated or unsaturated alkyl radical'.

It is not unusual to have the possibility that two variable substituents can join together to form a chain which is 'optionally unsaturated'. It is not enough to assume that this means optionally monoethylenically unsaturated unless there is a clear teaching in the specification that this is the intention of the applicant. If no such teaching exists, then the disclosure includes acetylenic unsaturation and, where appropriate, multiple unsaturation and this must be reflected in the database record.

The meaning of the phrase 'optionally substituted' may appear imprecise to a chemist but it is usually fully comprehensible to an attorney or to an experienced patent searcher. The common phrase 'phenyl optionally substituted by halogen' which may appear in the Markush formula as:-

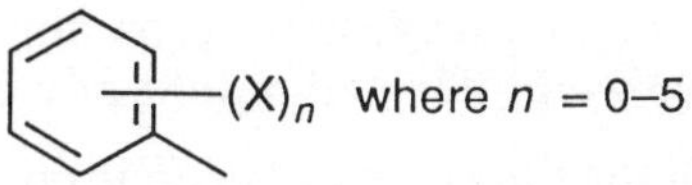

Figure 1

means that the phenyl ring may be unsubstituted or be substituted by one, two, three, four or five halogen atoms, and that the halogen atoms may be the same or different. It does not matter whether the examples contain only a mono-chlorophenyl ring; the examples exemplify the invention, they do not limit it. Only a clear statement in the specification that the number or type of substituent is restricted can change this. For example, you may find Markush formulae which contain 'a C1-4 alkyl group, optionally substituted by 1-3 halogen atoms'.

Similar restricting phrases are:-

1 'The term halo refers to Cl, Br and F'.

2 'The term cycloalkyl refers to saturated rings of 4 to 7 carbon atoms'.

3 'The term heterocyclo refers to fully saturated or unsaturated rings of 5 to 6 atoms containing one or two O or S atoms and/or one to four N atoms provided the total number of hetero atoms in the ring is four or less and bicyclic rings wherein the five or six membered ring containing O, S and N atoms as defined above is fused to a benzene ring'.

There has been some debate about the need to distinguish between what would appear to be identical disclosures, such as (1) R = halogen, and (2) R = F, Cl, Br or I, and about the necessity to be able to search exclusively for, say tert-butyl possibly together with suitable fairly specific generics such as tert-alkyl or branched chain alkyl, without retrieving references restricted to other specific alkyl groups such as methyl, ethyl, iso-propyl, etc., or to the less specific generic 'alkyl'.

These are very interesting from an academic point of view but hold little attraction for those who search within patent departments. If an attorney asks whether a compound having tert-butyl as a substituent at a particular point is novel, not only will he need to know whether this is the case but, in addition, he will want to know about all other compounds in which that substituent is either another specific alkyl group or the generic 'alkyl', especially if such compounds have been used for the same application. Only in this way, will he be able to draft a sound patent application, or give well informed patent advice to a client.

It would be a pity if the development of an input and search system giving such a high level of refinement delayed the introduction of a really good Markush database or gave one which was too expensive to use. Many of the queries which are thought to require this level of specificity should be searchable on the Registry File although this may require an extension of the selection rules for this database.

We are all aware of the so-called Derwent nasties which have been discussed elsewhere. These are not going to go away and there are going to be some patent documents in which the Markush formulae contain so many variables, each with such an extensive list of alternatives that the disclosure cannot be completely included in the database. It is just not possible with todays technology to cover every conceivable nuance of a very broad Markush formula. It is essential that the users of a Markush database have a list of all the documents which have not been covered fully in the database so that they can extend the search manually through these specifications if necessary.

I can illustrate my worries by reference to an American patent US 4822403. This is concerned with novel sulphonylurea derivatives which have the general formula:-

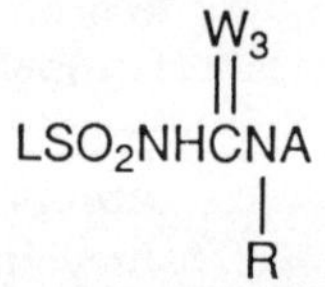

$$LSO_2NH\overset{\overset{W_3}{\|}}{C}\underset{\underset{R}{|}}{N}A$$

Figure 2

The specification is fifty seven pages long and of these, forty one are devoted to tables containing the structures of some 2650 compounds which fall within the general formula. Of these, physical data, in the form of melting points, are provided for about sixty individual compounds. Claim 1 contains twenty four variable groups some of which are stacked one within another whilst two can be linked together to form any one of four ring structures.

In the claims, A is restricted to a 1,3,5-triazine ring but in the specification A can be chosen from six distinct heterocyclic rings of which only two feature in most of the tables. Users will need to be assured that the database contains the full disclosure, not just the pyrimidine or triazine rings which feature in most of the tables, or the triazine ring of the claims?

In this and other sulphonylurea specifications, ring A can be a substituted 2,4,6-triazinyl ring as shown below:-

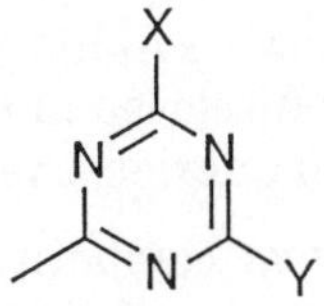

Figure 3

Various alternatives are given for both X and Y. For example, X can be a C1-4 haloalkylthio group and Y can be a ring structure of formula:-

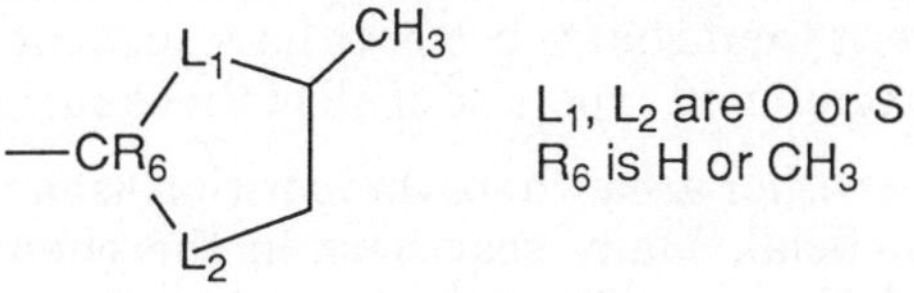

Figure 4

L can be any one of five different ring systems including phenyl, pyridyl and thienyl. These can be further substituted by a great variety of different groups.

For example, L can be a 2-pyridyl ring substituted at position 3 by a C2 alkynyl group which in turn can be substituted by phenyl ring optionally substituted by halogen, -CH3, -OCH3, -SCH3, -NO2. In addition, the pyridyl ring may be substituted on any of the three remaining ring carbon atoms by a variety of groups including an optionally substituted amino-sulphonyl group in which the substituents can, together with the nitrogen atom, form a morpholine ring.

Now, it is my contention that the test of a Markush database is whether or not it is possible to retrieve from the database hits in which the match occurs at the periphery of the database record. If this cannot be done, then there seems to be little advantage over the STN Registry File.

Therefore, both of the following search queries, in which R can be any substituent, should retrieve US 4822403.

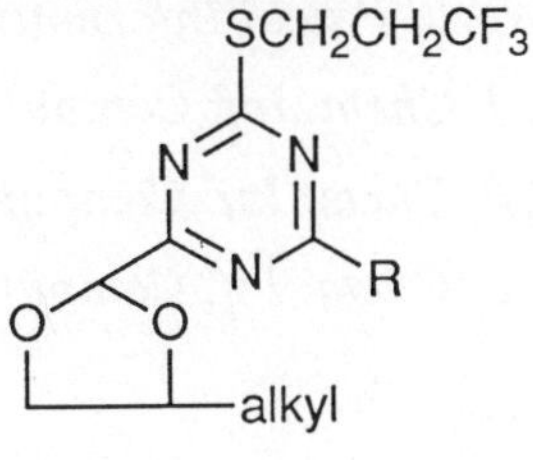

Figure 5

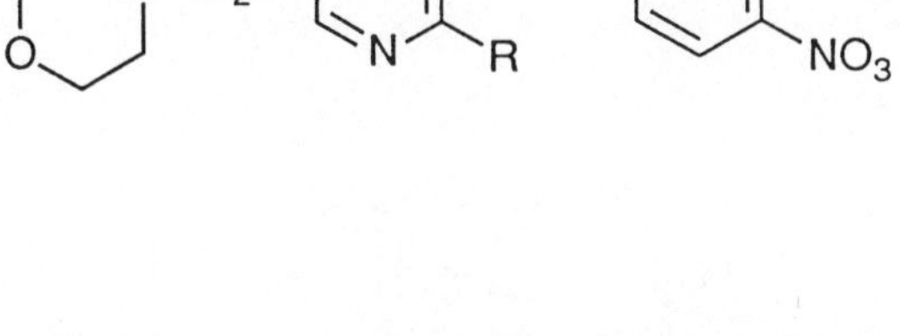

Figure 6

Neither query includes the sulphonylurea group which is an essential feature of the invention disclosed in US 4822403 but, if the record is not retrieved by such search queries, then the database has failed.

Whilst I believe firmly in competition, the production of a Markush database seems to be one area in which it would be better to have co-operation between the two major producers and also with Professor Lynch from Sheffield University. There is no point in developing two or more Markush databases which nearly meet the needs of the few potential users; it would be far better to have just one joint database benefiting from the combined expertise of all of those currently working in the field.

There is another area where competition seems to be detrimental rather than beneficial. Many searchers in the chemical area, who view the Chemical Abstracts file as being complementary to WPI, would like to see the Derwent files accessible through STN, which is their preferred host. This could lead to fruitful cross-fertilisation with the Chemical Abstracts Registry Numbers and indexing being available for the WPI file. A similar thing has happened on Orbit, with the merging of the American Petroleum Institute APIPAT file with WPI. This, incidentally, adds some Registry Numbers to the WPI file.

The market for patent information is limited and it is a pity that of the new services we have discussed today, only EPOQUE and a really good Markush database could bring significant benefits to those working in the area. Of these, one will not be made available to those who could make most use of it and the other seems likely to fail unless the producers of the current databases either co-operate or undertake to support databases which will almost certainly lose money.

References

[1]. *World Patent Information*, **12**(1), 55, 1990.

[2]. K H Pilny, *European Intellectual Property Review*, 183-184, 1991.

[3]. S M Kabak, *World Patent Information*, **12**(1), 45-46, 1990.

[4]. K A Cloutier, *J. Chem. Inf. Comput. Sci.*, **31**(1), 40-44, 1991.

[5]. N R Schmuff, *J. Chem. Inf. Comput. Sci.*, **31**(1), 53-59, 1991.

[6]. J M Barnard, *J. Chem. Inf. Comput. Sci.*, **31**(1), 64-68, 1991.

Patent statistics: comparing grapes and watermelons

Edlyn S. Simmons

Marion Merrell Dow Inc. Cincinnati, Ohio 45215-6300 USA

Nancy Lambert

Chevron Research & Technology Company, Richmond, CA 94802-0627 USA

Ever since patent databases were introduced, we have been hearing that patent data are a terrific source of competitive intelligence. Ever since statistical software has become widely available, we have been hearing how much you can learn from patent statistics. However, two groups with very different backgrounds are writing about patent statistics. Statisticians tell us about the wonderful trends you can discover with patent statistics [1]. Patent experts warn us that we should be careful when we analyze patents statistically not to compare apples and oranges [2]. Statistical studies assume that the differences among the members of a set of items are small enough to overlook. This is generally true when large sets of items are being analyzed; the physical chemist can easily overlook differences in properties in a set of 6.02×10^{23} molecules. But patent statistics seldom involve Avogadro's number of patent documents.

Statistics work best on large sets of uniformly sized items. These produce small standard deviations, so reliable conclusions can be drawn from the data. But as the size of the set of data being analyzed is reduced or the variability of the properties of the items in it grows, the conclusions become unreliable. Apples and oranges are both fruits, and about the same size, but they are not the same thing. In the field of patent statistics, we often analyze sets of a few hundred patents from more than one country. In this case, we are not merely comparing apples and oranges, we are comparing grapes and watermelons. They are both fruits, but that is the only thing they have in common.

How patents differ

Patents can vary in any number of ways, even when they are issued in the same country. They can vary in the type of invention claimed, in their scope, and in the commercial value of the invention. A patent can claim a chemical substance *per se* or a mixture of substances; a process for preparing a compound or a method for using it; a method for performing a manufacturing operation or the product of the operation; a machine, a method for performing an operation on the machine, or the product it produces — a whole fruit bowl of different types of inventions. The claims can specify a single compound or a large genus of compounds, any one of which is protected by the patent. Some patents cover a multitude of products. A patent that claims a new chemical substance covers all products that contain the substance and all processes that employ it. In that sense, such a patent can be compared with a watermelon: It is big enough to serve a crowd. On the other hand, some patents cover only a limited number of products. A patent that claims an improvement in a known process can protect only the improved process and cannot be used to prevent others from using the unimproved process or from devising a different improvement to achieve the same end result. That kind of patent can be compared with a single grape. A product can be covered by a group of patents that protect various aspects of the product. For instance, a new catalyst useful in petroleum processing may be covered by a basic patent on the composition and a whole host of related patents — sometimes all based on the same priority — covering different uses. This kind of patent protection can be compared to a bunch of grapes: Even if none of the individual patents is substantial in its own right, their cumulative coverage can be quite broad.

Patent statistics are used to monitor the level of innovative activity in a company, to discover the areas of technology under research and changes in research direction, and to evaluate the relative importance of the research being conducted at various companies and even in various countries. There is no question that much of a qualitative nature can be learned by studying the numbers and kinds of patents being published by various research-based organizations. However, the focus of statistics is quantitative, and many publications on patent statistics have emphasized the quantitative evaluations that can be made by analyzing patents without considering how much the patents may vary.

Patents vary considerably in their economic value. The value of a patent depends upon the market for the claimed invention. Pharmaceutical companies file thousands of patents claiming new compounds every year, but only a small percentage of these produce revenue; the number of new drugs entering the market every year is very small. When a new drug is approved, both the patentee and other companies file patent applications claiming methods for making and administering it. Many of these companies are never licensed to make or sell the approved drug, and many of the processes are never used; so these patents may have no commercial value. In short, the patent portfolio of most major companies resembles a bowl of fruit with a melon or two, a lot of apples and oranges, and bunches of grapes. There will also be a large box of raisins: patents that have expired, lapsed, or been invalidated.

VARIATION IN PATENT LAW AND PROCEDURE AMONG USA, EPO, JAPAN AND USSR

COUNTRY:	USA	EUROPEAN PATENT (1-14 Countries)	JAPAN	USSR
PATENT TERM:	17 YEARS FROM GRANT	20 YEARS FROM FILING	15 YEARS FROM GRANT; MAXIMUM 20 YEARS FROM FILING	15 YEARS FROM FILING
SEQUENCE OF PATENTING PROCEDURES:	PRIOR ART SEARCH & EXAMINATION GRANT/PUBLICATION	PRIOR ART SEARCH PUBLICATION EXAMINATION GRANT/PUBLICATION NATIONAL VALIDATION/ OPPOSITION	PUBLICATION (KOKAI) PRIOR ART SEARCH & EXAMINATION PUBLICATION (KOHOKU) OPPOSITION GRANT	PRIOR ART SEARCH & EXAMINATION GRANT/PUBLICATION
ALTERNATIVE PATENT COVERAGE:		NATIONAL PATENTS IN MEMBER COUNTRIES		INVENTOR'S CERTIFICATES FOR SOVIET CITIZENS
CHEMICAL SUBSTANCES *PER SE*:	PATENTABLE BROAD GENERIC CLAIMS ACCEPTED	PATENTABLE BROAD GENERIC CLAIMS ACCEPTED	PATENTABLE ONLY SINCE 1976 NARROW GENERIC CLAIMS ACCEPTED	COVERED ONLY IN INVENTORS' CERTIFICATES

Figure 1: Variation in patent law and procedure among USA, EPO, Japan and USSR

How patenting philosophies differ

Patents also vary because of differences in patenting philosophy. These can be differences in policy between companies, they can result from differences in the need for patent protection from one industry to another, and they can reflect differences in national patent law.

Between companies

In Europe and North America some companies have corporate policies that encourage the filing of patent applications on all possible innovations, while other companies file patent applications only when major profits are expected from the innovation. The number of patent applications filed by a company can change radically without any changes in the level of research simply because a new senior-level executive believes that more patent applications should be filed or that, on the other hand, corporate resources are being wasted by filing too many patent applications.

Between industries

Industries differ in their patent needs. New products in some industries can be developed for market in months and become obsolete within a few years. Companies in those industries are likely to file patent applications quickly and abandon the applications on obsolete inventions without waiting for a patent to be granted. Other industries need years of development time before they can tell whether they will be able to market an invention. Some inventions can be sold everywhere in the world, while others have a limited geographical market; and the patentees will have very different foreign filing philosophies. In the pharmaceutical industry, for example, potential patients exist all around the world, and a product takes years to get through the regulatory system and onto the market; so a great many patent applications are filed internationally and maintained for years, just in case. In the oil industry, relatively few countries have oil fields, so there is no point in filing foreign patent applications on drilling technology everywhere in the world.

Between countries

Different countries' patent laws reflect their varying philosophies. Some of the differences among the patent laws of the US, Japan, the USSR, and the EPO are shown in Figure 1. We would expect differences between countries with strong patent systems and those with weak patent systems; but even among countries with strong patent systems, substantial philosophical differences exist. Americans and Europeans think of patents chiefly as offensive tools to protect potential products and as defensive tools to block the competition from getting patents that would cover similar products. Only a granted patent provides protection, but published unexamined patent applications are prior art that prevents competitors from getting patents. Companies based in the United States always file patent applications with the intention of obtaining a patent. Without publication of the unexamined application, the US patent application's defensive power is extremely limited. US law accepts patents with broad generic claims and with combinations of claims of various types, so that large research programs can be covered by a single patent application.

Both the law and the philosophy of patents in Japan are very different. Until the late 1970s Japanese patents were limited to a single, narrowly defined inventive concept, and chemical substances could be protected only by process claims; so many patent applications had to be filed to cover a research program. In this sense, you can compare Japanese patent applications with a bunch of grapes: There are a lot of them, each one bite-sized. Even though the law has been modified, the Japanese philosophy of patenting encourages the filing of large numbers of patent applications of narrow scope. Companies based in Japan file a great many patent applications with no intention of having the applications examined and obtaining a patent. We are told that Japanese companies publish the results of their research in patent applications as much to share information as to block other companies from patenting the same inventions. The fact that the patentees consider the bulk of these applications to be of little commercial value is demonstrated by the small fraction that are ever claimed for priority on equivalent patents filed in other countries or published as examined patents. Statistical analyses that fail to differentiate between the publication of Japanese patent applications of narrow scope and of granted United States patents of broad scope are truly comparing grapes and watermelons.

Country code & postings	Result of GET analysis, Kind of Document Code (KD)		Meaning of Kind of Document Code (KD)
EP (9657)	THERE ARE 7 UNIQUE VALUES		
	2781	A1	PUBLISHED APPLICATION
	2605	A3	*Search Report*
	1977	A2	PUBLISHED APPLICATION, NO SEARCH REPORT
	1784	B1	**Patent**
	444	A4	*Supplementary Search Report*
	35	B2	Revised patent
	31	TD	German translation of patent claims
US (6715)	THERE ARE 4 UNIQUE VALUES		
	6653	A	**PATENT**
	44	E	Reissue patent
	10	H1	*Statutory Invention registration*
	8	B1	Reexamination certificate
JP (28478)	THERE ARE 4 UNIQUE VALUES		
	21228	A2	PUBLISHED APPLICATION
	6600	B4	**Examined application**
	478	T2	Translation of PCT application
	172	X2	Unpublished document
SU (4860)	THERE ARE 3 UNIQUE VALUES		
	4500	A1	**INVENTOR'S CERTIFICATE**
	277	A2	*Addition to Inventor's certificate*
	83	A3	**PATENT**

FIRST PUBLICATION OF SPECIFICATION
First publication of accepted claims
Non-patent document

Figure 2: December, 1990, EPO, US, Japanese and USSR patent publications, INPADOC postings

European patent applications are a lot like oranges, because up to 14 separate units of patent protection are supplied under a single skin. Not many trivial patent applications are filed in the EPO. A patentee that wanted protection only in its home country would file for a national patent application, so all European patent applications may be considered foreign-filed. Thus applicants for European patents are more likely to request examination, and most patentable inventions eventually result in patents.

In the Soviet Union, as might be expected in a Communist country, little economic incentive has existed in the past to protect inventions with patents. A new patent law is under consideration at the time this is being written, but the law in effect at the beginning of 1991 provides for Inventor's Certificates that acknowledge authorship of an invention made by a Soviet citizen, with rights to use the invention vested in state enterprises. Although many inventor's certificates are issued, very few of these are granted to non-Soviet citizens. In December 1990, only 8 of 4500 claimed foreign priority. Although patents are also granted in the Soviet Union, very few are actually issued.

International patent databases

Patents vary in the way they are reported in the patent databases. Patent offices issue a lot of different kinds of documents, and the major international patent databases index more than one kind of document from the countries that publish more than one kind. The nature of the documents differs from country to country. Figure 2 shows some of the kinds of documents issued in a typical month, December 1990, by the patent offices of the United States, Japan, the Soviet Union, and the European Patent Organization. The patents were retrieved from the INPADOC database on ORBIT, and analyzed for the Kind of Document code with the GET command. You can see a substantial difference in the number of documents published by the four offices and in the number of different kinds of documents. Some of these are granted patents, of course. Japan and the EPO publish patent applications that have not been examined or granted. These will be published a second time if a patent is granted, generating a second record in INPADOC. The United States and the USSR have no pre-examination publications. In the Soviet Union, most of the documents are not really patents, but rather inventor's certificates that give the state the right to practice the invention. If the number of postings alone is compared, it would appear that Japan and the EPO issued many more patents than the United States issued, and the USSR issued nearly 5000. When the number of granted or examined patents is compared, however, we can see that Japan and the United States issued nearly identical numbers of patents, while the EPO granted far fewer and the USSR granted very few indeed.

Many of the patent documents that showed up in the analysis are not even colorable imitations of patents. The EPO issues search reports that are simply part of the patent examination procedure. If you were to count each of these documents without regard to their legal effect, you would end up with vastly distorted statistics. When you analyze these documents you must be careful to compare the granted apples only with other granted apples and not with the published application oranges. And you

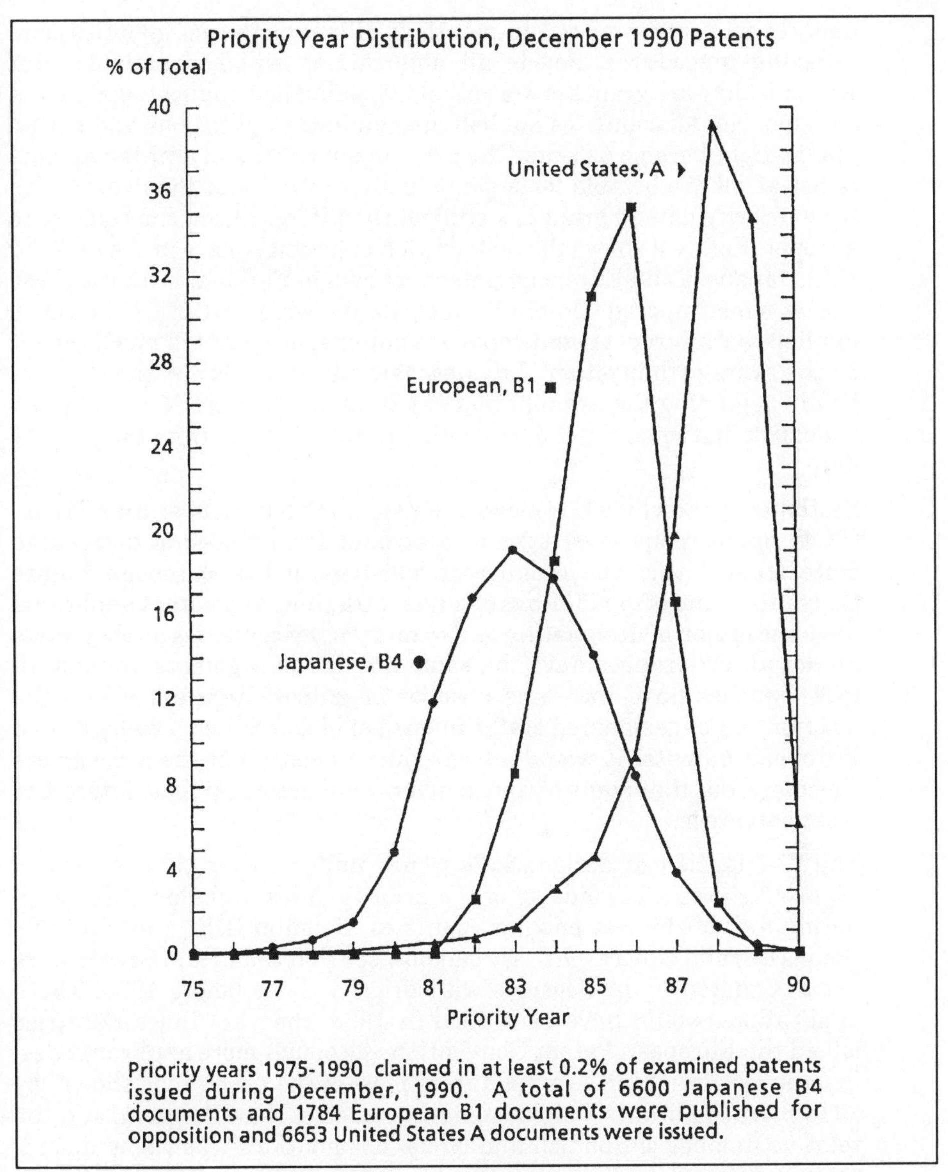

Priority years 1975-1990 claimed in at least 0.2% of examined patents issued during December, 1990. A total of 6600 Japanese B4 documents and 1784 European B1 documents were published for opposition and 6653 United States A documents were issued.

Figure 3

must be extremely careful not to count the documents that are neither apples nor oranges.

One of the problems with tracking patenting activity over time is establishing the time frame. This is not a serious problem if the objective is to track the publication of unexamined patent applications in countries that publish them 18 months after filing. The priority date can be tracked and published documents will be available 18 months later; with a slight delay to account for missed 18-month deadlines by the patent offices and indexing procedures, nearly all applications will be available. But whether they are grapes or watermelons, published applications are not yet ripe. Not all countries publish unexamined applications and not all applications become patents. The delay in publication of granted patents is not at all predictable for a single patent office, and the average lag from priority date to grant is significantly different from one country to another. Figure 3 shows the distribution of priority years in the granted U.S., Japanese and European patents shown in Figure 2, with the three curves superimposed. United States patents, which are much slower to publish than European and Japanese patents, are granted much sooner on the average than either. Japanese law allows a delay of up to 7 years before beginning the examination of patent applications, resulting in a wide distribution of grant dates with a peak at 7 years from the priority date.

Statistics that include European patents and Patent Cooperation Treaty (PCT) applications must take into account the number of designated states they cover. The number of members of the European Patent Convention and of the PCT has changed with time, and patent applicants have the option of designating as few or as many countries as they wish. So not all EP oranges have the same number of segments and not all PCT bunches have the same number of grapes. Figure 4 shows the distribution of designated states in our set of December, 1990, granted European patents. It would appear that almost all of them designate Germany, but that many of the member countries are of little interest to most patentees.

The distribution of designations is not uniform over time, however. Figure 5 shows a breakdown of the priority dates of the patents designating four of the less popular countries, Belgium (BE), Austria (AT), Spain (ES) and Greece (GR). We can now see that Spain and Greece were not designated in applications with priority dates before 1985. Those applications would have been filed in 1986, the year those countries joined the European Patent Convention. Although more applicants designated Belgium than Austria during most of the time range shown, by 1988 more applicants were designating Austria than Belgium, and the relative number of Spanish and Greek designations was rising disproportionately. So statistics generated from the international patent databases can be misleading to someone not 'skilled in the art' who attempts to interpret them.

Patenting companies — grapes or watermelons?

One of the best uses of patent statistics is to compare the patent holdings and patenting patterns of different companies. Who files the most patent applications at home and who files the most equivalent patents in foreign countries? Which companies are getting the most patents in your com-

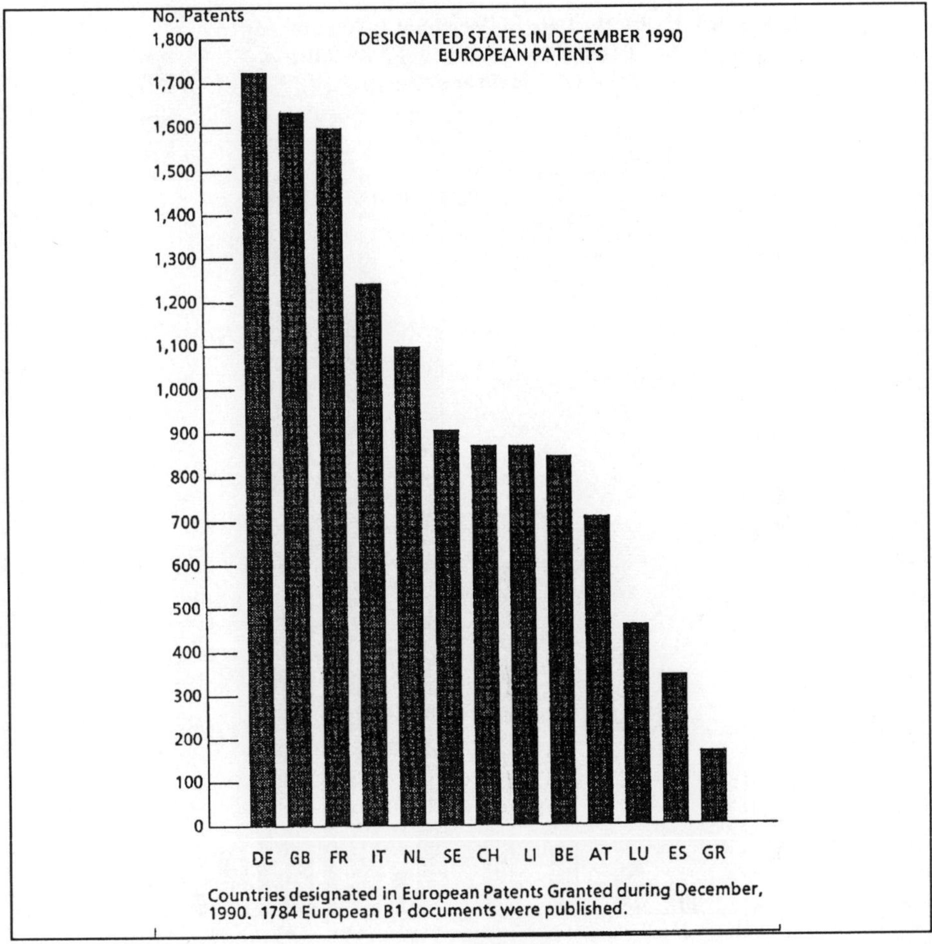

Figure 4

pany's technology? Have new patentees emerged as major competitors in the field? Have the major competitors changed the focus of their research? You can learn those things by counting the number of patents assigned to various companies and analyzing them to find out what they are about, when they were filed, and where in the world equivalent patents are being published.

However, the corporate world also includes grapes and watermelons, and they can cause major distortions in statistical analyses. Some companies are monolithic organizations that assign all of their patent applications to the parent company. Other major companies are clusters of separately named divisions, subsidiaries, and related companies of which the parent company has partial ownership. Figure 6 shows two different counts of the number of US patents issued to major oil and chemical companies in

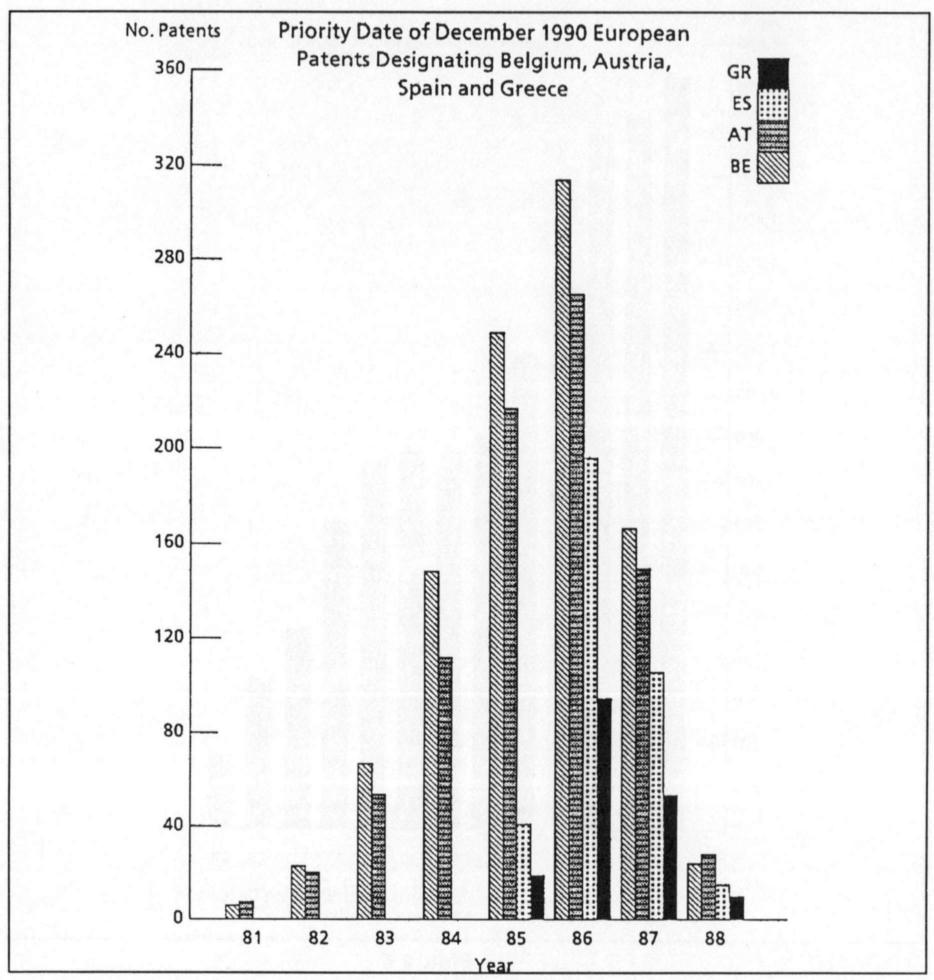

Figure 5

1990. The gray lines on the bar chart show the figures supplied by the United States Patent and Trademark Office for the Intellectual Property Owners and published in the *IPO Washington Brief* list of the top 200 Corporations Receiving US Patents in 1990. The black bars show the number of patents assigned to the same companies, but this time patents assigned to divisions, subsidiaries and partially owned companies have been added to the patents assigned to the parent company. It is pretty clear from this graph that the relative positions of some companies is switched if the subsidiaries are included. In fact, subsidiaries of some of these companies are listed separately in the IPO list.

The Patent and Trademark Office presumably counted the patent assignees in its internal assignment files, but several sources of patent

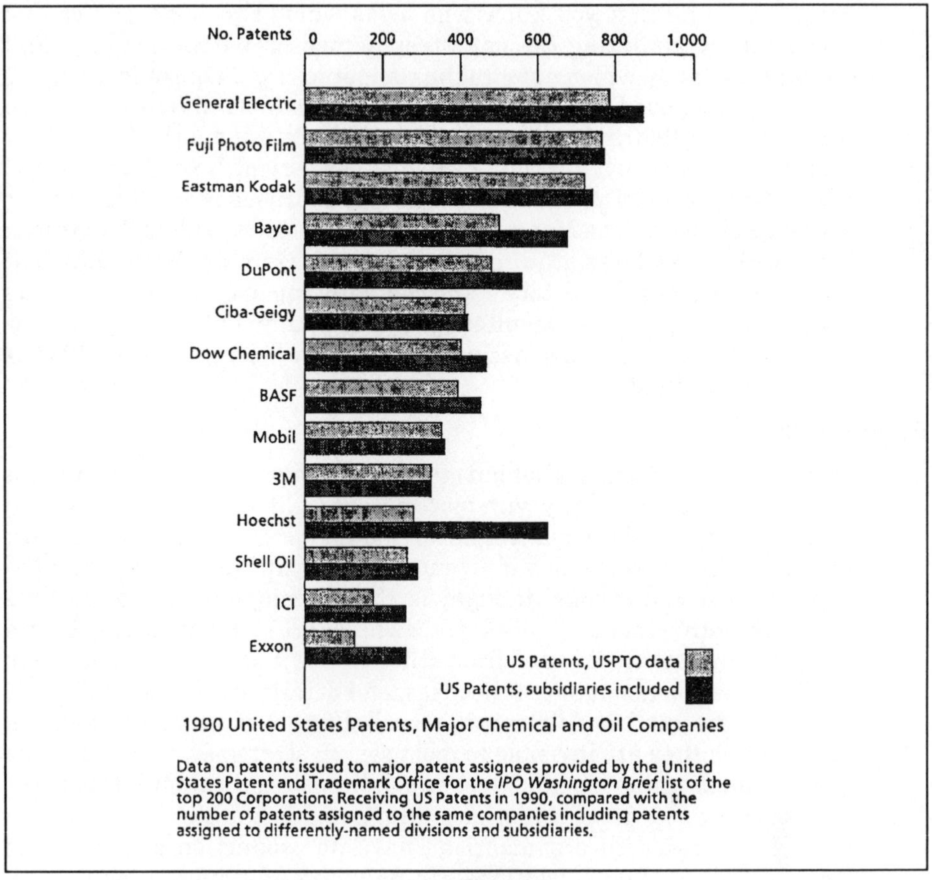

Figure 6

assignees are available on online information services. The Derwent US Patents file indexes the data provided to it by the USPTO. To produce the World Patents Index database, Derwent compares the assignee names with a thesaurus of standard company names and company codes. Many, but not all, subsidiaries are given the company code of the parent company, and some irregularities in the form of a company name are corrected. Assignee names are indexed from all of the patents in a patent family; so when members of the family are assigned to different corporate entities, all of the assignees are in the record. IFI/Plenum has a thesaurus of standard company names and company codes for all US chemical companies and larger companies in other technologies. Like Derwent, IFI applies the standard codes to some but not all subsidiaries. INPADOC uses a thesaurus of patentee names, but it corrects only the form of a company name, not the name itself. To get reliable data on who owns what, you really have to know who owns whom.

Even assuming that you know who owns whom this year, you cannot really tell by analyzing the patentees named on patents or in patent databases exactly which patents the companies own. One of this paper's authors is employed by a company called Marion Merrell Dow Inc., which was formed in 1990 by combining Marion Laboratories with Merrell Dow Pharmaceuticals Inc, a wholly owned subsidiary of Dow Chemical Co. Until 1981 Merrell Dow was a division of Richardson Merrell Inc. Other divisions of Richardson Merrell formed a new company called Richardson Vick, which was later acquired by Procter & Gamble. With sufficient information about the history of the company you can search for patents that were assigned to progenitors of the company, but finding out how many of those patents are assigned to Marion Merrell Dow in 1991 is much more difficult.

Patent licensing

One of the basic premises behind statistical analyses of corporate patent holdings is that a company with more patents will have better protection from competition in the marketplace. And it is certainly true that patent protection from competition is an enormous advantage. What the statistical approach overlooks, though, is that a company can protect its products with patents it does not own. A great many products are protected by patents licensed from others. Patent databases do not have any information about patent licenses, and details are difficult to find in the rest of the published literature as well. But patents that are licensed, even though they are invisible to most search strategies, are among the most valuable patents of all. To be worth licensing, a patent has to cover a product, or at least a potential product, and it has to be valid and in force. Some research organizations have no production or marketing facilities; all of their inventions are patented and offered for license. Some production and marketing organizations have no research and development departments; all of their products are sold under patent licenses. Statistical analysis of patentee data credits the patent owner for the patent, of course. But the benefit of the patent in the marketplace is shared by the licensee, who gets no recognition at all.

Diltiazem

Some of the points discussed above are illustrated by the patent history of the antihypertensive agent diltiazem. Until recent years, most Japanese pharmaceutical companies had no subsidiaries in the US or Europe and offered licenses on their new products to companies who could market the products in other countries. One of those products, diltiazem, was licensed by the innovator company, Tanabe, under patents based on a priority application filed in Japan in 1968. Diltiazem has been an extremely successful product, and a statistical analysis of patents covering diltiazem filed from 1968 to 1988 illustrates the fact that patent statistics can be used to track the progress of a technology

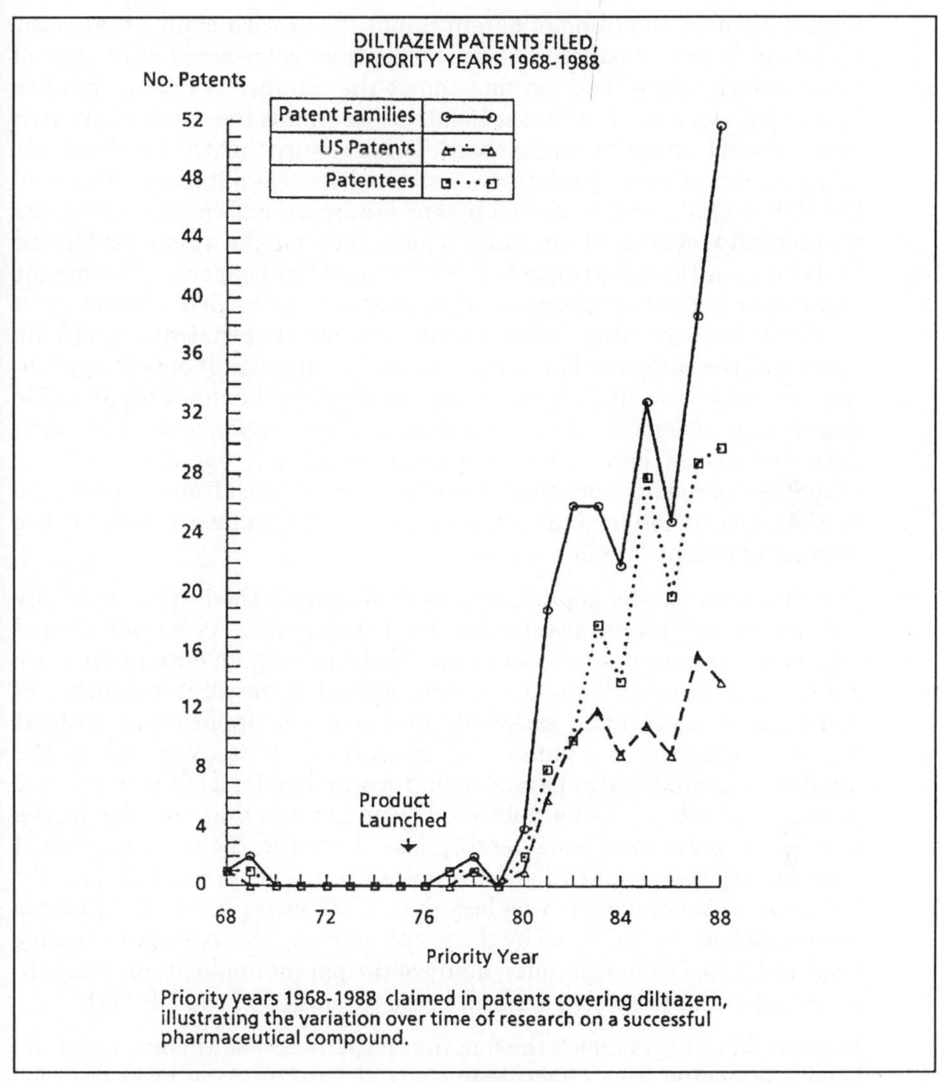

Priority years 1968-1988 claimed in patents covering diltiazem, illustrating the variation over time of research on a successful pharmaceutical compound.

Figure 7

as it emerges and matures. This kind of analysis assumes that the number of patents being issued correlates with the amount of research activity and that research will result in the publication of patents. Published studies of this type have concluded that some technologies are better suited to this type of analysis than others. Where the number of patent publications actually does correlate with the amount of research activity — and in the pharmaceutical industry it probably does — the results can be dramatic.

Figure 7 tracks the filing of patent publications with claims that cover diltiazem in some fashion over time. The time is tracked by the year of latest priority date. The top line shows the number of patent families indexed by Derwent, but the set of data is not simply the result of a search of the World Patents Index database. To make sure that the set of patents being analyzed was complete, we searched several databases and read the claims of a member of each patent family to confirm that diltiazem was actually covered. The number of patentees and the number of United States patents are also tracked. It is not surprising that only a few patent applications covering diltiazem were filed before the first marketing of the drug. After the drug had achieved some market penetration, and the success of the compound became evident, the number of patent applications began to rise, and an increasing number of companies began to file patent applications. As the expiration date of the original patent covering diltiazem *per se* approaches in various countries, patents are being sought by more and more companies who were barred from entering the marketplace by the original diltiazem patents. So a sharp increase in the number of patent families occurs.

The fact that patent applications are being published does not really mean that patentable inventions are being made. As we mentioned before, many patent applications are filed claiming inventions that are not really new, useful, and nonobvious enough to result in patents; and many patent applicants, especially in Japan, file applications without any expectation that a patent will be granted. When you look at the number of United States patents being granted instead of the number of patent applications being published, you can see that the rise in the number of patents is considerably less dramatic. These are granted patents; rejected applications are not shown, and the cutoff date for inclusion in this analysis was less than 2-1/2 years before the patents were collected. As indicated by the graph showing the average pendency time for United States patents, many of the patent applications published elsewhere in the world are still being examined in the USPTO.

You should never overlook the fact that patents do not all have the same legal and competitive effect. Hundreds of patents have been filed for diltiazem formulations and processes; but the first patents to be filed in most countries, the ones based on the Tanabe 1968 application, claim the compound *per se.* In most countries those patents are real watermelons: They have prevented nearly all of the later patents belonging to companies without a license under Tanabe's patent from being exploited by their owners. Exceptions, strangely enough, are the Japanese patents. At the time Tanabe filed its application on diltiazem, Japanese law permitted claims to chemical compounds only in the form of process claims. It should not be surprising, therefore, that many Japanese companies have filed patent applications claiming other processes for making diltiazem. Most of the later patents are very narrow in scope:

They are grapes. These patents cover a single improvement in a manufacturing method, or the treatment of a new symptom, or a combination of diltiazem with a second drug or a new excipient; and each one will cover only those products that use the improved process or the combination of ingredients.

What can we learn about diltiazem from these data? We can see that, as the original patent on diltiazem approaches its expiration date, an increasing number of pharmaceutical companies hope to sell some form of diltiazem as a generic drug. We could infer from the level of interest that diltiazem is a very successful drug, although it would have been less trouble to learn that directly from sales figures. If we read the patents instead of merely counting them we can learn even more; we can see which companies are interested in selling diltiazem, where they hope to sell it, and what kinds of products they are interested in selling.

US and international patent classification changes

Changes in the patent classification schemes over time can cause serious distortions in analyses of technical trends in industry and of changes in emphasis within a single company. They can mask actual changes in research emphasis and create the appearance of changes that have not occurred. The original purpose of classification codes is to facilitate the examination of pending patent applications by patent examiners. A patent classification system divides all of the technologies in patents into subsets and assigns one or more classification codes to each patent or other document in the examiners' search files, allowing the prior art to be searched without wasted effort. But the patent system itself, in its charter to promote the progress of science, guarantees that the technology being patented will change with time. When new technologies are introduced, the patent offices must create new classifications for them. When an area of technology becomes more specialized or when an old class accumulates too many documents to be used for a convenient search, the existing classifications have to be revised.

The United States patent classification system is under constant review, and groups of patents are reclassified whenever a revision seems to be warranted. The patents shelved in the search rooms are given new classification codes, and new classifications are recorded in computerized records. The trends shown by analyzing a set of patents according to their US patent classification codes will be different if the current classification codes are used rather than those shown on the face of the printed patent. These differences are reflected in the various patent databases available online. Most databases, including the Derwent US Patents databases, record the information printed on the face of the patents at the time of issue (or maybe the information on the USPTO tapes at the time the database was originally loaded); but the CLAIMS databases are reloaded every year with current classification data.

The dramatic differences these approaches create can be judged from the results of a GET analysis of patents issued to ALZA Corp. ALZA is a US pharmaceutical company that specializes in drug delivery systems. Figure 8 shows the results of parallel searches in ORBIT files USPM and CLAIMS, on a day when both databases were updated to July 2, 1991. Only 2 ALZA patents were in the CLAIMS database and not in USPM, but the distribution of classes assigned to the patents appears to be radically different in the two databases. If you look at the second and third pairs of bars in the graph, you can see that Class 128 is assigned to 180 of the ALZA patents in USPM, but only 68 of the same patents as indexed in CLAIMS; while class 604 appears on only 166 of the USPM records and on 237 of the CLAIMS records. Both classes deal with medical devices, and that entire area has been reclassified by the USPTO. A similar reclassification shows in the comparison of classes 424 and 514, both of which relate to pharmaceutical formulations.

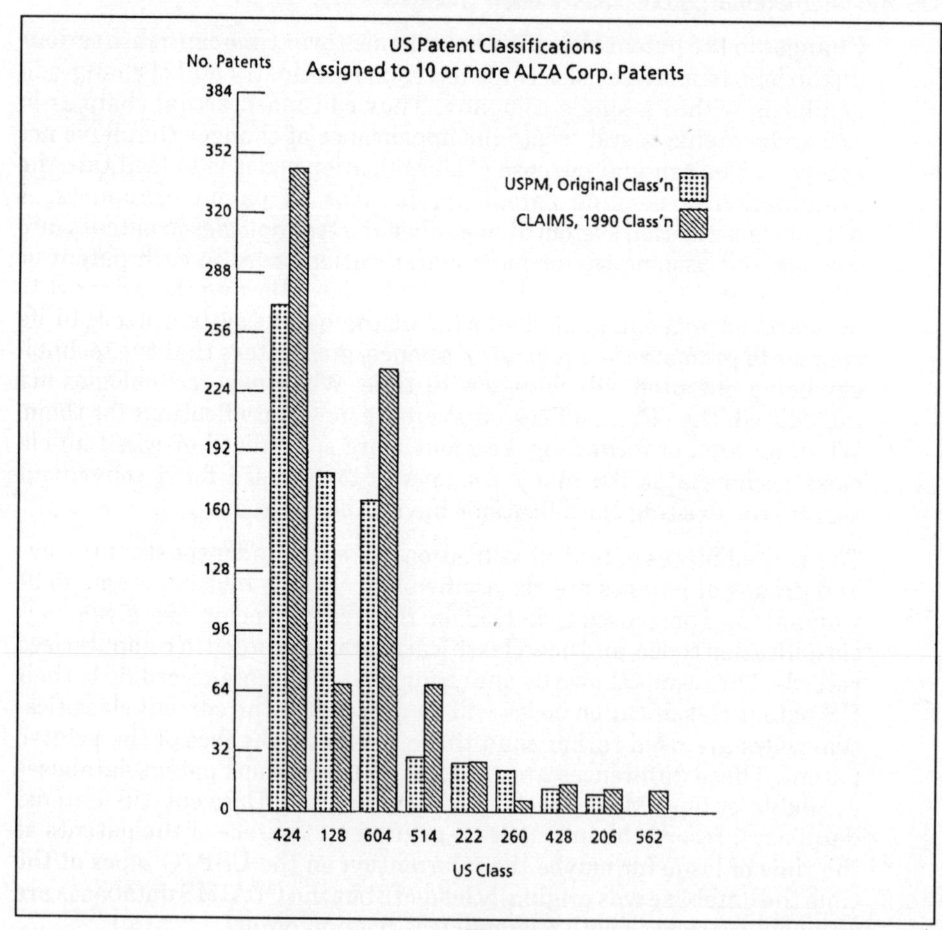

Figure 8

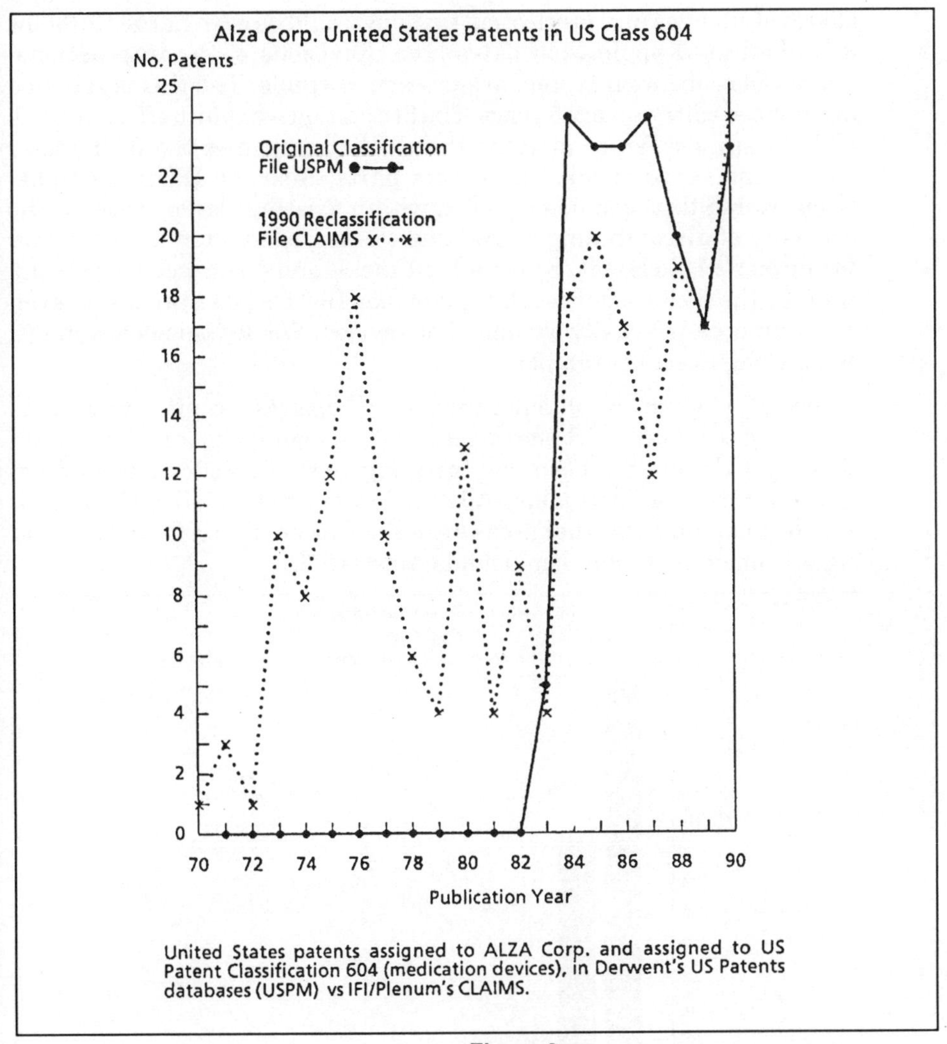

United States patents assigned to ALZA Corp. and assigned to US Patent Classification 604 (medication devices), in Derwent's US Patents databases (USPM) vs IFI/Plenum's CLAIMS.

Figure 9

The differences between the original classes and the 1990 reclassification data are even more apparent when we analyze the assignment of ALZA patents to US class 604 over time. It would appear from the data graphed in Figure 9 that the USPTO established class 604 in 1983 and subdivided it in 1988. The reclassifications may have been an attempt to accommodate the very inventions ALZA was patenting. Rather than the classification system showing how ALZA's research was changing, we may be seeing how ALZA's research was changing the classification system.

Although it is usually more meaningful to analyze trends in research by means of patent application or priority dates rather than issue dates, the effects of changes in classification systems would be even harder to follow if we had used application dates here. Revisions of the International Patent Classification system are more predictable. The IPC is reissued in a revised edition every 5 years. The 5th edition went into effect in 1989. IPC classes are larger in scope than US classes, and the 3-character classes can be subdivided into more parts. Also the IPCs are alpha-numeric and thus can be expanded within existing classes without the necessity of discontinuing or creating classes, as is often necessary for the numeric US classes. So usually at the 3- and 4-character level, and often at the 6-character level, the broader IPC classes stay intact even when an area is heavily expanded or revised. We see this below in our genetic engineering example.

Figure 10 is a bar graph showing the IPC classes assigned to patents belonging to another US pharmaceutical company, Upjohn. These data show that Upjohn has been concentrating most of its efforts on medical and veterinary science, Class A61, and organic chemistry, Class C07, with less emphasis in other areas including biochemistry, Class C12, and organic macromolecular compounds, Class C08.

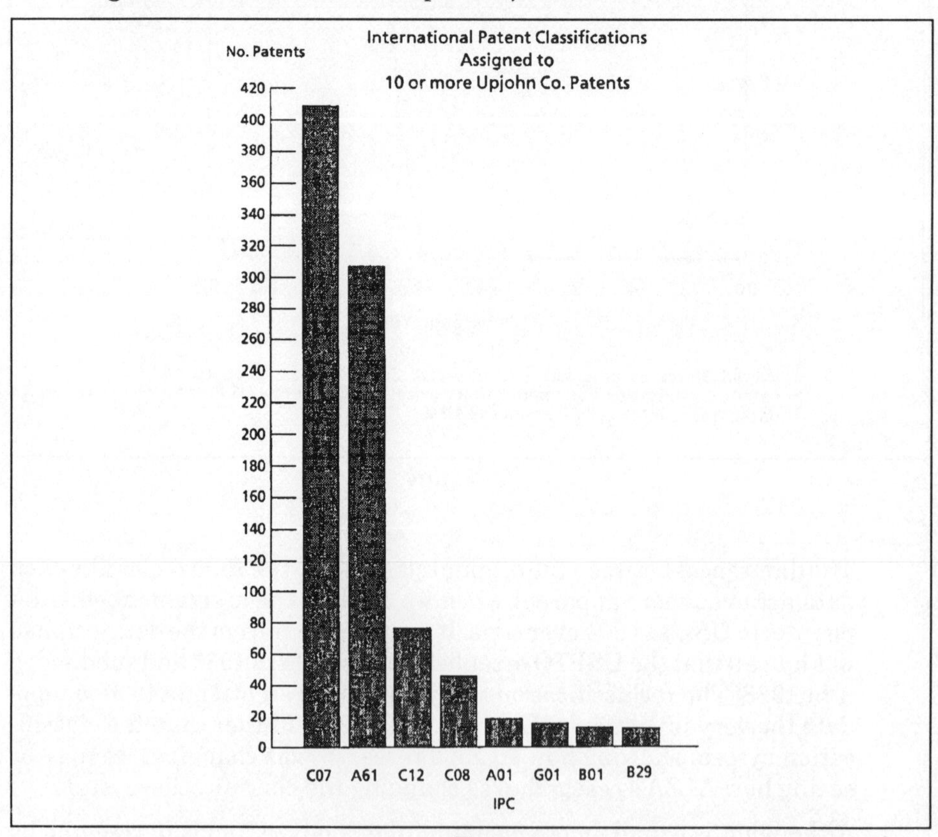

Figure 10

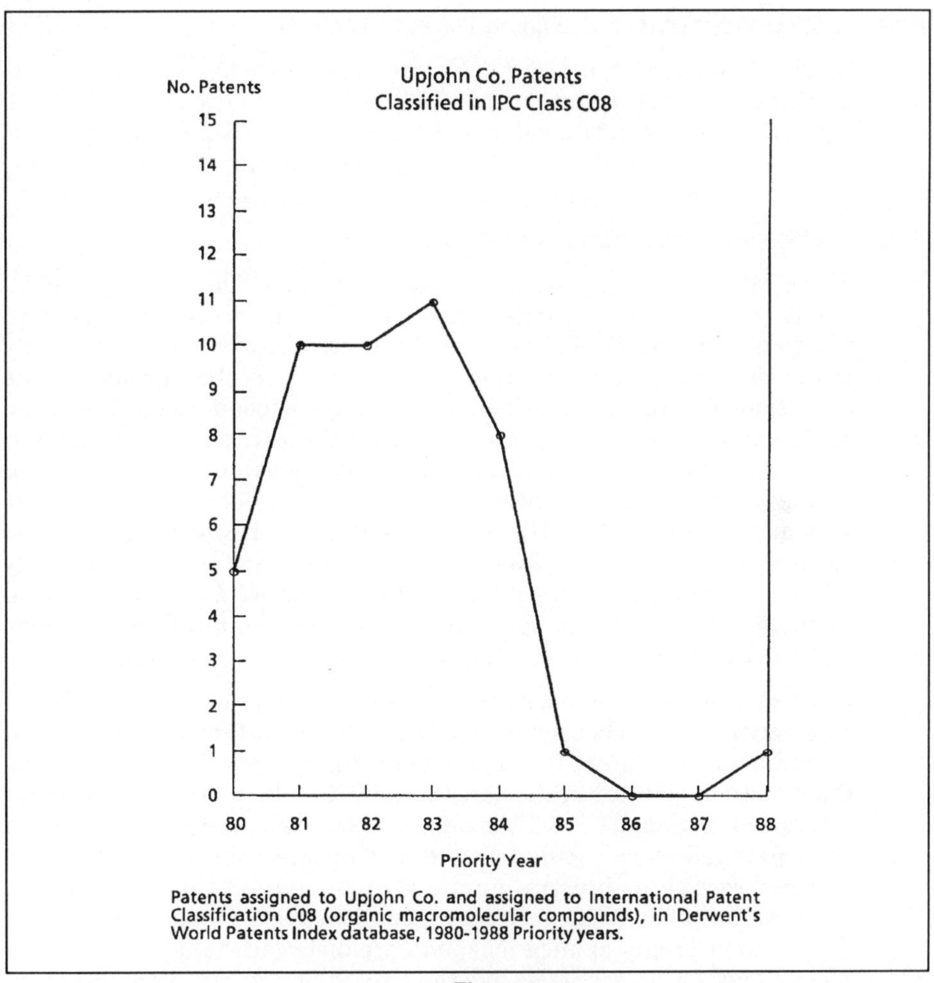

Figure 11

That last class is a little unusual for a pharmaceutical company. When we took a closer look at the distribution of Upjohn patents in Class C08 over time in Figure 11, this time using the priority date rather than publication year, we discovered that after a fairly consistent level of activity in the early 1980s, Upjohn filed only two Class C08 patents. If we have learned much from our analysis of ALZA's patents in US class 604, we will check to see whether IPC C08 was abolished when the 4th edition of the IPC went into effect.

Instead, however, this change is the result of a real shift in Upjohn research: Upjohn had sold its polymer division.

Patent citation statistics (see also figure 11A at end of text)

Much of the recent interest in patent statistics has focused on the usefulness of patent citation statistics for the evaluation of the technology in the patents and the relationships between cited and citing patents. Such studies overlook a number of variables in the patent examining process and must invariably end up comparing apples and oranges.

Patent examiners' varying citation practices

Three different sources exist for patent citations. The one required by law is the examiner's search of the prior art. The second source, often mentioned in studies of citation analyses, is citations made in the body of the patent in the description of prior art or the background of the invention. The third source of citations is correspondence between the applicant and the examiner during prosecution of the patent, which can be an informal report of prior art cited by other patent offices (required of applicants in most countries) or the formal information disclosure statement mandated by US patent regulations. However, there is no requirement that the patentee have done a search of the prior art, only that he or she report on the nearest known prior art. Citations provided by applicants are not automatically included as examiner citations. Some examiners choose to include them all; others are very selective.

As this suggests, examiners vary in their citation practices. They follow only general rules about what kinds of references they should cite. All examiners are required to cite any publications they find that describe the invention claimed in the patent application they are searching; but individual examiners have considerable freedom to decide whether a particular reference should be cited in a rejection for obviousness and whether references illustrating the general state of the art should be cited. All three kinds of citations are published in European and Patent Cooperation Treaty applications, and are distinguishable from one another. However, in countries like the United States, the patent application is not published at all if the examiner can cite references that teach or suggest the invention that was claimed. So 'killer references' — the best references, those that read on the claimed invention — never appear in a published patent and thus never show up in patent citation studies. Nor are citations relevant to the patentability of the claimed invention differentiated in US patents from those providing background of the technology.

Patent examiners are not restricted in the kind of references they cite. Scientific journals are searched in fields where they are likely to yield citable references. Patent examiners also frequently cite reference books. A search of the US Patents database tells us that *Advanced Organic Chemistry: Reactions, Mechanisms, and Structure*, a textbook by Jerry March, has been cited in 531 patents, and *Remington's Pharmaceutical Sciences* has been cited in 216. These publications are long and detailed summaries of the prior art, extraordinarily useful as sources of examiners' citations. However, since the search files in patent offices are based in their own published patent specifications, these patents are cited most frequently.

A look at recent heavily-cited US patents yielded some very interesting information about examiner citation practices. We neighbored on the CIT (number of citations) field in the CLAIMS file on ORBIT to produce a list of US patents issued since 1982 which had subsequently been cited more than 90 times. There were only two. One was listed as being cited 97 times; the other, 100. These patents turned out both to be to the same company (Union Carbide) on similar technology, and shared two inventors (Flanigen and Lok):

```
-1-
PN    - US4440871
TI    - CRYSTALLINE SILICOALUMINOPHOSPHATES
IN    - CANNAN THOMAS R; FLANIGEN EDITH M; GAJEK RICHARD T; LOK
        BRENT M; MESSINA CELESTE A; PATTON ROBERT L
PA    - UNION CARBIDE CORP (87136)
PD    - 84.04.03
AP    - 82.07.26 82US-400438
PCL   - 502214000, CROSS REFS: 208114000, 208136000, 208138000, 208213000,
        208254000H, 585418000, 585467000, 585475000, 585481000, 585482000,
        585489000, 585660000, 585670000, 585740000
CIT   - Cited in 100 patents

-2-
PN    - US4310440
TI    - CRYSTALLINE METALLOPHOSPHATE COMPOSITIONS (IN THE
        PRESENCE OF AN ORGANIC TEMPLATING AGENT)
IN    - FLANIGEN EDITH M; LOK BRENT M; WILSON STEPHEN T
PA    - UNION CARBIDE CORP (87136)
PD    - 82.01.12
AP    - 80.07.07 80US-166333
PCL   - 502208000, CROSS REFS: 208112000, 208114000, 208135000, 208136000,
        208138000, 208143000, 208213000, 208254000H, 280114000, 423305000,
        502510000, 502511000, 585418000, 585419000, 585467000, 585475000,
        585481000
CIT   - Cited in 97 patents
```

Note that both are classified in US class 502 (catalyst compositions) and cross-referenced in classes 208 (petroleum processing) and 585 (hydrocarbon compositions); the older one is also cross-referenced in class 423 (inorganic chemistry). Are these patents technologically important? Perhaps — but that may have nothing to do with their frequency of citation.

More to the point, they are both very long patents. The more recent one runs for 78 columns and has 38 claims and 54 very detailed examples of the relevant catalyst preparations. The older one is 52 columns long, with 26 claims and 64 examples. What a wealth of detail for an examiner to cite! In the USP file, we searched these patent numbers as cited patents and produced for each the set of more recent patents citing it. These numbered 135 and 126, respectively (since the Number of Citations field in CLAIMS is not as up to date as the bibliographic data in US patents databases). Interestingly, the sum of the sets was only 173 patents. In other words, the majority of patents citing one of these also cited the other. So the rest of this analysis is on the combined set of US patents citing either of these two patents.

Next we determined who — which patent examiners — were citing these two very heavily cited patents. An analysis on the Name of Examiner (NOE) field for the 173 citing patents proved most enlightening.

NO. OF PATENTS	EXAMINER NAME
45	DOLL, JOHN
40	BRENEMAN, R. BRUCE
26	MCFARLANE, ANTHONY
15	DEES, CARL F.
11	PAL, ASOK
9	DAVIS, CURTIS R.
8	JOHNSON, LANCE
6	LEEDS, JACKSON
6	SNEED, H. M. S.

It would appear that nearly half the citations were made by only two examiners, the top two listed. When we investigated further, we found that John Doll is a supervisory examiner whose name appears along with those of less experienced examiners on many patents. However, R. Bruce Breneman has been examining only since the mid- '80s (the first patents on which he is listed as an examiner came out in 1987); only 189 patents in the USP file list him as an examiner, and John Doll is listed on 128 of these.

In which of the 189 patents that Breneman has examined does he cite these two heavily-cited patents? That is, in which classes do they fall? When we search Breneman's name directly as an examiner against the 173 citing patents, we retrieve 42 patents. (This is more than the 40 that showed in the analysis above. Two more listed slight variations on his name as examiner — the sort of name-scatter that can introduce error into name analyses.) When we analyze these 42 for their patent classifications (both original and cross-reference), we generate the following list:

NO. OF PATENTS	PATENT CLASSIFICATION
39	423/PCL
22	502/PCL
6	208/PCL
2	585/PCL
1	425/PCL

In fact, all 42 of the patents in which Breneman cited our heavily-cited patents fell into one of the first three classes listed — logically enough, since all three of these classes appear on our two heavily-cited patents, and examiners tend to examine, classify, and cite in the patent classes of their own art units. In turn, 134 of the 189 patents that Breneman has examined fall into these three classes. So, out of only 134 patents, the ones which Breneman has classified into classes 423, 502, or 208, he has cited one or both of our highly-cited patents in 42 of them — nearly one-third.

Next we took a look at Breneman's citation habits in general. We did an analysis of the CT (citations) field of his 189 patents and determined that he cites, on average, 10.8 patents in each patent that he examines. A look at a random-over-time 60 of his 189 patents gave the same figure and showed that he also cites, on average, 2 literature references per patent. Given this very high average, we wondered if Breneman had other patents besides our two heavily-cited patents that he tended to cite frequently. So, finally, we analyzed the 189 patents that he has examined and ranked by frequency of citation the patents he cited in them. Predictably, he has cited a large variety of patents — 1250 separate ones altogether. But our two heavily-cited patents are by no means his only 'pet' patents. They lead the list, but Breneman has a total of 15 patents which he has cited 14 or more times each in the six years or so that he has been examining.

NO. TIMES CITED	PATENT NUMBER
39	US4310440
37	US4440871
28	US4500651
28	US4567029
27	EP--55529
25	EP--59059
24	EP--54364
24	EP--55046
23	US4420467
18	US3941871
18	US4554143
17	US4061724

16	US4456582
16	US4486397
14	US3702886
9	GB2024790
9	US4208305
8	US4385994

This analysis shows that US patent examiner R. Bruce Breneman tends to cite patents generously, and that he tends to keep 'pet patents' handy which he cites very frequently indeed. Nor, we suspect, is he unique among patent examiners in these tendencies.

To illustrate this, next we looked for the most highly-cited US patents of all. Two patents were listed in CLAIMS as having been cited 267 times each. These are as follows:

-1-
PN - US3702886
TI - CRYSTALLINE ZEOLITE ZSM-5 AND METHOD OF PREPARING THE SAME
IN - ARGAUER ROBERT J; LANDOLT GEORGE R
PA - MOBIL CORP (56432)
PD - 72.11.14
AP - 69.10.10 69US-865472;
CONTINUATION IN PART OF: 67.04.14 67US-630993 (Abandoned)
EQ - DE1792783; FR1564164; FR1587860; FR1587861; GB1161974
PCL - 423328000, CROSS REFS: 208111000, 423277000, 423326000, 502060000, 502061000, 502077000, 502202000, 556173000
CIT - Cited in 267 patents

-2-
PN - US2825721
TI - POLYMERS AND PRODUCTION THEREOF
IN - BANKS ROBERT L; HOGAN JOHN PAUL
PA - PHILLIPS PETROLEUM CO (65688)
PD - 58.03.04
PCL - 526106000, CROSS REFS: 260DIG025, 273DIG004, 273008000, 422134000, 502256000, 526101000, 526102000, 526335000, 526340200, 526348000, 526348200, 526348300, 526348400, 526348500, 526348600, 526348700, 526351000, 526352000, 528493000, 528498000, 585530000
CIT - Cited in 267 patents

The older patent, like the two Union Carbide patents, is very long, with many claims and detailed examples. The more recent one is interesting: It is the first Mobil patent on ZSM-5 zeolite, an extremely common petroleum-processing catalyst.

The Mobil patent is recent enough — 1972 — to allow some further investigation in the USP file, since most of its citing patents (all but the ones from the missing 1971-1974 data) will be in USP. Again we produced the set of citing patents (numbering 286 in USP) and ranked their examiners, with the following results:

NO. OF PATENTS	EXAMINER NAME
63	MEROS, EDWARD J.
41	GANTZ, DELBERT E.
29	DOLL, JOHN
29	SCHMITKONS, G. E.
23	DEES, CARL F.
21	DAVIS, CURTIS R.

This time the top examiner, who is not a supervisory examiner, accounts for over 20% of the times this patent was cited. Breneman is clearly not the only examiner with 'pet' patents.

Also of interest: Of the 286 citing patents, nearly half — 139 — were other Mobil patents. This was quite predictable. Mobil is a company known for patenting abundant clusters of patents in its technologies, in which other members of the same cluster would be cited frequently.

Citation comparisons between US patents with large and small patent families

What do we find if we look, not at individual examiners' citation habits, but at citation patterns for related groups of US patents? Here patent statistics can shed some light on, and sometimes contradict, other patent statistics.

Patent citation analyses most often analyze US patents cited in other US patents, and many advocates of patent citation statistics claim that the most heavily cited patents are, by definition, the most technologically important patents. These analysts do not equate technological importance with commercial importance, but they are rather fuzzy when asked to define technological importance. We can assume that patent owners base foreign filing decisions on the relationship between the technical advance claimed in a patent and the commercial advantage that will result if competitors are prevented from taking advantage of it. If we permit the owners of patents to define the patents' commercial importance in terms of how much their owners will spend to obtain them, we can reasonably conclude that patents filed in a greater-than-average number of countries are considered important.

If we divide a set of US patents into a subset of inventions with large families and another with small families, we can compare the average number of times the patents in each subset have been cited, to see if any significant differences show up. Something vitally important to this process is to ensure that patents in both subsets cover the same time range; since, all other factors equal, younger patents will of course be cited less frequently than older patents — they have not been around, and available to be cited, for as long.

The comparison of large-family and small-family patents becomes rather complex, because it cannot be done online with any accuracy. ORBIT does permit ranking on the NP (number of patents) field, which is a count of how many individual patent numbers appear in the PN (patent number) field. However, the NP field does not include countries listed separately as priority countries or designated states (the latter especially important in an analysis of country coverage); it does count more than once the same country listed more than once (for instance, in different publication stages). A more accurate ranking is available via PatStat (PatStat Plus Version 2, a PC statistical analysis software package from Derwent Publications Ltd.), which permits a 'patent families' analysis of a set of downloaded patents. This analysis lists individual patent families in the set being analyzed (referencing the 'basic' patent, the first family member that Derwent picked up) ranked by their total country coverage: the sum of the separate countries that show up in the patent number, designated states, and priority fields.

The table produced by PatStat (see below) lists each class of country separately, followed by a total ('full range') that eliminates duplicate countries. The 'full range' is the closest measure available of all the countries in which protection has been sought for that patent. The only countries missing from the families being analyzed — and this affects current patents more than older ones — are non-priority countries in which the patent has been filed but not yet published.

JUNE 19, 1984: U.S. PATENTS WITH LARGE FAMILIES

Basic patent	PN family	PN range	DS range	PR range	Full range
WO8401068-A	8	6	28	1	32
EP--92807-A	19	18	9	1	26
WO8202246-A	6	3	21	1	23
EP--84948-A	16	12	11	1	22
US4455239-A	14	11	11	1	22
EP--73060-A	11	10	11	1	21
WO8300015-A	19	13	16	1	21
EP--24701-A	13	11	10	1	20
NL7606825-A	25	20	-	1	20
EP--71121-A	12	10	10	1	19
GB2071094-A	32	19	-	1	19
WO8102126-A	9	7	18	1	19
EP--70362-A	12	10	8	1	18
EP--73741-A	11	10	9	1	18

EP--87968-A	12	10	9	2	18
WO8103603-A	13	11	14	1	18
WO8302715-A	6	4	15	3	18
BE-877733-A	16	16	-	1	17
BE-892831-A	18	17	-	1	17
BE-895054-A	17	17	-	1	17
BE-895055-A	19	17	-	2	17
DE3202919-A	8	7	11	1	17
EP--11473-A	23	11	7	1	17
EP--79004-A	11	9	9	1	17
US4455383-A	9	8	9	1	17
WO8302469-A	12	8	14	1	17

To divide a set of patents into small- and large-family, we must first perform a search on Derwent. To avoid both time and subject biases, we started the search in the CLAIMS database, and chose as a set all US patents issued on 19 June 1984. This produced 1175 patents. These we transferred via their patent numbers to the Derwent World Patents index. Here we created a subset consisting of patents with EP or WO family members and patents with an NP of 5 or more, to reduce the size of the set to be downloaded. We reviewed the table created by the PatStat analysis and eliminated patents with a 'full range' of fewer than five. On the remaining patents we performed the necessary word processing to produce an ASCII file of patents numbers that could be uploaded to ORBIT. We uploaded the numbers and searched them on Derwent. This became the set of wide-country-coverage patents. We then reproduced the original search and subtracted the uploaded patents from the total. This became the set of small-family patents.

The only citation ranking available is on the CLAIMS file, which includes a field enumerating how many times individual US patents have been cited. This field is GETtable. So, the next step was to transfer our two subsets back into the CLAIMS file. This we did by transferring and searching priority application numbers (since the Derwent patent families contained too many individual patent numbers to fit within ORBIT's PRINT SELECT limits). However, this process created a time bias by introducing additional US patents with the same priorities as the original set (continuations, divisionals, and so on). The large-family subset contained more of these patents not in the original set of 19 June 1984 patents (an additional 40%, compared to an additional 33% for the small-family set); and these new patents were, on average, younger than the original set. Hence the time bias. So the final step was to limit the two subsets transferred into CLAIMS to the original set of 19 June 1984 patents.

For each subset the GET command was performed on the CIT (number of citations) field. This resulted in listings of numbers of citations per patent, each next to the number of patents in the set having that many citations. Multiplying numbers of patents times frequency of citation, and adding the products, gave the total times all patents in the set were cited, as reported in the CLAIMS database. Dividing that by the number of patents in the set (including those with zero citations, not shown in the listings) gave the average number of citations per patent.

The large-family US patents totaled 381.

No. patents	Cite/pat.	No. cites
89	1	89
47	2	94
41	3	123
33	4	132
14	5	70
10	6	60
14	7	98
6	8	48
2	9	18
5	10	50
2	11	22
3	12	36
1	13	13
2	14	28
3	15	45
1	16	16
1	17	17
2	20	40

Total no. cites for set		999

Average cites/patent = 999/381 = 2.62

The small-family US patents totaled 683.

No. patents	Cites/pat	No. cites
149	1	149
90	2	180
63	3	189
56	4	224
29	5	145
21	6	126
17	7	119
6	8	48
10	9	90

8	10	80
4	11	44
2	12	24
4	13	52
1	14	14
3	15	34
1	18	18
1	24	24
1	25	25
1	26	26

Total no. cites for set		1622

Average cites/patent = 1622/683 = 2.37

Our statistical consultant found the differences between the averages to be statistically insignificant [3]. In other words, in a one-week sample of US patents, the number of cites per patent is essentially the same for the subgroup of commercially important patents — those with a country coverage of 5 or more — as for the subgroup of patents with a country coverage of fewer than 5. As shown below, this may be explained, at least in part, by the possibility that equivalent patents are being cited — a phenomenon that would, of course, affect large-family patents more than small-family patents. But it is individual patents, not families, that are being analysed in most patent citation studies.

Citation comparisons between active and lapsed US patents

We did a separate analysis of patents whose commercial importance was defined by another variable: whether they had been allowed to lapse for non-payment of maintenance fees. US patents filed after December, 1980, have been subject to maintenance fees on their 3rd, 7th, and 11th anniversaries. If these fees are not paid, the patents lapse one year later. So most US patents granted between 1982 and mid-1987 have gone through at least one round of pay-the-fee-or-lapse. We did a search similar to the one described above, but with a larger set of patents: all those granted in June 1984. This produced a set of 5502 US patents. These we divided into the subsets of lapsed (768 patents) and still active (4734 patents), taking advantage of the fact that this status is searchable directly in CLAIMS. On these subsets we performed the same citation analysis as we described above, and came up with average-citations-per-patent figures of 1.72 for the lapsed patents and 2.46 for the active ones — a difference that our statistical consultant found to be significant [4]. Many of the active patents, having been filed before maintenance fees were required, may actually belong to the set of patents less valued by their owners, so the correlation may be a bit stronger than these figures indicate.

Unfortunately, this correlation between the number of citations and the value of patents cannot be used to identify individual valuable patents. Most important, the statistical correlation refers only to average numbers of citations; there are many patents in the active set that have never been cited and many in the lapsed set have been cited often (the only three patents cited more than 20 times were in this set). Furthermore, patent examiners cite only one member of a patent family, so the number of patents citing an individual patent may differ radically from the number of patents citing the technical information disclosed in each equivalent patent.

In this analysis and the one before, we have not disproved the theory that the more important patents are being cited more frequently; but we have shown flaws in the methodolgy being used to measure frequency of citations. The following discusses and illustrates one of these flaws in more detail.

Citations of patent families

Keep in mind that the citations being analyzed above, like those in most published studies of patent citation rates, were only of individual patent numbers, and only those cited in US patents. One of the difficulties with such an approach is that a patent is not a unique document. Many patent applications, especially those that are considered to be very valuable by their owners, are filed in more than one country and form a cluster of interchangeable documents. Quite different citation figures result from analysis of full patent families, and analysis of citations made in EPO and PCT search reports as well as US patents.

Examiners cite references for what they teach, and it makes very little difference which member of a family of patents is cited. Counting each family member as a separate patent can distort the results of a statistical analysis to the extent that relative numbers of citations are meaningless; but that is all that most citation analyses are capable of doing.

To illustrate this point, we studied the number of citations to members of four patent families. We found two US patents and two European patent applications that had been cited 11 or more times during 1990 in US patents in US patent class 514. We searched these in the full Derwent file, WPAT, to determine the size of their patent families and to compare US citation rates for the individual patent numbers, for all US members of the patent families, and for the full patent families. We also looked at how often these patent families had been cited in European and PCT patents (the only other countries' citations currently searchable online).

One of the patents had only one family member, a US patent; and it had been cited in 5 EPO and PCT patents. Citations of the other three are shown in Figure 12.

CITATION PATTERNS, 3 DERWENT FAMILIES OF PATENTS CITED HEAVILY BY THE USPTO

Patent Number	US4507140	EP--40345	EP-170006
Total US citations	40	46	23
Derwent family members	8	45	11
US citations of all family member	69	69	23
US citations of US family member(s)	40 cites (1 of 3 US)	24 cites (3 of 8 US)	0 cites (1 US)
EP/PCT citations of all family members	37	35	18

Figure 12

Items of interest to note: All three patents had one or more US family members, of which only one of three, three of eight, and none of one, were cited at all in the US. Two of the three patents had significantly higher citation rates for their whole families than for any individual family member. All three families were heavily cited in EPO and PCT patents, but the corresponding European patent application was the family member cited in nearly all cases.

If it were actually true that heavily cited patents were technologically more important than others, how would we explain so many equivalent patents without citations disclosing the same technology as the heavily cited patents?

A look at rare-country patents

The patents that a company considers important ought to include, not just those filed in many countries, but particularly those filed in countries where that company does not usually apply. These exceptional patents might just be real plums! Such patents can be found online through statistical analysis of patent countries. Such an analysis will not include EPO or PCT designated countries. EPO designated countries can be considered 'common' and thus of little interest for this analysis, but the PCT includes many third world countries. These countries are not indexed by Derwent or Chemical Abstracts except as designated states, and some are not even indexed by INPADOC. For companies that frequently file patent applications through the PCT, and for the exceptional PCT applications filed by others, a different analysis will be necessary.

First let us look at Exxon's patents from 1984 through 1986 — ones old enough to have essentially complete patent families but recent enough that most of Derwent's current 'minor' countries are being covered. We searched for Exxon patents with an accession year between 1984 and 1986 and, ignoring PCT applications for this example, did a GET analysis on the first two characters of the Patent Number field. This produced a ranked list of patenting countries. We did the analysis to a 'select' list so that we could search the countries directly.

Sel. #	No. patents	Patenting country
1	857	US/PN
2	444	EP/PN
3	407	CA/PN
4	285	J6/PN
5	266	DE/PN
....		
23	4	PT/PN
24	4	SE/PN
25	2	HU/PN
26	1	J/PN
27	1	CH/PN
28	1	CS/PN
29	1	J8/PN

We requalified the codes to search as patenting countries and selected the 'real' countries (eliminating variations on Japan) with the fewest postings: Hungary, Switzerland, and Czechoslovakia. We printed a few hits.

-1-

AN - 86-341127/52

TI - Emulsifiable oils for use in metal-working oils and hydraulic fluids - having hard water compatibility etc. contain combination of alkanolamine and water-sol. hydroxy-di: or tri:carboxylic acid

PA - (ESSO) EXXON CHEM PAT INC

IN - LENACK ALP,KECH FJ

PN - EP-206833-A 86.12.30 (8652)
J62018496-A 87.01.27 (8709) {JP}
BR8602966-A 87.02.17 (8712)
AU8659274-A 87.01.08 (8714)
ZA8604678-A 87.08.27 (8747)

CN8604443-A 87.03.11 (8822)
HUT046054-A 88.09.28 (8843)
ES2000177-A 88.01.01 (8914)
US4956110-A 90.09.11 (9039)
DS - AT BE CH DE FR GB IT LI LU NL SE
PR - 85.09.16 85GB-022841 85.06.27 85GB-016301
AP - 86.06.26 86EP-304964 86.06.27 86JP-151324 86.06.23 86ZA-004678
86.06.27 86ES-000022 88.04.12 88US-180436

-2-
AN - 86-341090/52
TI - Olefin polymerisation catalyst - comprises metallocene and an alum-oxane
reacted together on support
PA - (ESSO) EXXON CHEM PAT INC
IN - WELBORN HC
PN - EP-206794-A 86.12.30 (8652)
AU8658914-A 86.12.24 (8706)
J61296008-A 86.12.26 (8706) {JP}
NO8602447-A 87.01.12 (8708)
DK8602924-A 86.12.22 (8711)
FI8602625-A 86.12.22 (8714)
BR8602880-A 87.03.17 (8721)
HUT042103-A 87.06.29 (8730)
ZA8604568-A 87.12.21 (8812)
ES8802395-A 88.08.16 (8839)
US4808561-A 89.02.28 (8911)
US4897455-A 90.01.30 (9012)
CA1268754-A 90.05.08 (9025)
CS8604580-A 90.07.12 (9037)
AU9170079-A 91.04.11 (9122)
DS - AT BE CH DE FR GB IT LI LU NL SE
PR - 85.06.21 85US-747615 88.03.18 88US-170485 88.12.02 88US-278910
AP - 86.06.23 86EP-304806 86.06.20 86JP-143142 86.06.19 86ZA-004568
86.06.20 86ES-556357 88.03.18 88US- 17048 88.12.02 88US-278910

Only someone within Exxon's patent department can verify whether these are the patents that Exxon considers most important or potentially profitable. They may simply be patents in technologies which happen to have potential markets in these 'rare' countries.

Second let us look at Upjohn's patenting practices. These produce some extremely interesting country coverage statistics. For example, where does Upjohn file patent applications for routine inventions and for exceptional ones, and how does their use of the multi-country patent applications introduced in the late '70s, the EPO and PCT patents, affect their national filings in other countries? To generate some of these

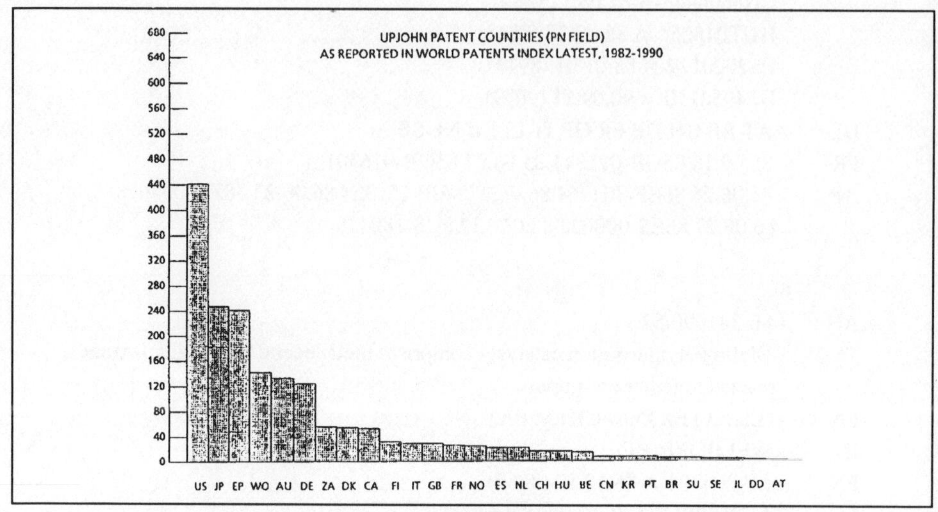

Figure 13

statistics, we produced a set of all Upjohn patents in Derwent which were first published (accession year) between 1982 and 1990 — 666 patents — and downloaded them for PatStat analysis. The distribution of patent countries, without regard to the filing dates, is shown in Figure 13. We did a cross-tabulation analysis of priority year *vs.* patenting country over 1981 through 1988, and produced the spreadsheet shown in Figure 14:

This analysis shows only individual patenting countries and reflects the peculiarities of individual countries' treatments of PCT and EPO patents (as well as how these are treated by Derwent). In other words, one needs a considerable knowledge of both international patent law and Derwent indexing policies to avoid some pitfalls in interpreting these data.

Let us take a closer look at Upjohn's patenting practices as they are reflected by the distribution of countries over time. PCT patents (shown as WO) have been available since 1979; but Upjohn only started applying for them in 1986. On the other hand, Upjohn has been applying for EPO patents at least since 1981, but appears to have begun tentatively. That is straightforward enough; but what can we say about different countries' patents covered by the EPO and PCT? The distributions for a few of the EPO member countries are shown graphically in Figure 15. If we look at the top graph, French applications stopped in 1983; but West German national applications seem to have continued until 1986. A few Spanish applications continued to be filed, until 1986. When we see the distribution of EPO filings on the lower graph, the picture changes. In reality German national applications probably stopped the same year as French national applications. When an EPO patent is granted, it is registered in West Germany and given a German (DE) patent number. These are

Priority Year Distribution, Upjohn Co. International Patents

Country Code (PN)	81	82	83	84	85	86	87	88	Total
AT	1								1
AU	4	3	6	5	5	27	35	49	134
BE	5	4	4	0	0	0	0	0	15
BR	4	0	0	0	0	0	1	6	6
CA	9	11	13	9	3	0	0	6	53
CH	11	3	1	0	0	0	0	0	17
CN	0	0	0	2	1	1	1	2	7
DD	2	0	1	0	0	0	0	0	3
DE	29	31	27	14	13	2	0	0	125
DK	4	3	2	4	5	10	14	12	55
EP	18	34	35	30	34	27	34	26	241
ES	6	4	4	3	4	0	1	0	22
FI	6	3	2	3	3	5	5	5	32
FR	17	4	1	0	0	0	0	0	27
GB	17	4	1	0	0	0	0	0	29
HU	6	2	2	2	1	1	1	1	16
IL	1	2	1	1	0	0	0	0	5
IT	15	6	4	1	0	0	0	0	30
JP	34	42	41	31	31	26	31	2	248
KR	1	1	3	1	1	0	0	0	7
NL	12	4	1	0	0	0	0	0	22
NO	5	2	1	1	2	5	4	5	25
PT	3	1	0	2	0	0	1	0	7
SE	3	1	0	0	0	0	0	0	5
SU	3	1	0	1	0	0	0	0	6
US	82	99	58	54	30	21	16	13	442
WO	0	0	0	0	24	30	38	50	142
ZA	7	17	13	6	6	2	2	2	57

Figure 14

showing up in Derwent's PN field, and in the spreadsheet they are indistinguishable from genuine German national patents. They stop in 1986; which is about right, since the average EP patent granting time is about 4 years, so that EPO patents applied for in 1986 and 1987 are just now being granted and registered in Germany. Since it has been possible to designate Spain in a European patent application since 1986, it is possible, but not certain, that Upjohn has not stopped filing for Spanish patent protection.

Similarly, when we look at a comparison of United States, Japanese, and Australian patent publications with and without PCT applications in Figure 16, we can explain some of the changes in country coverage.

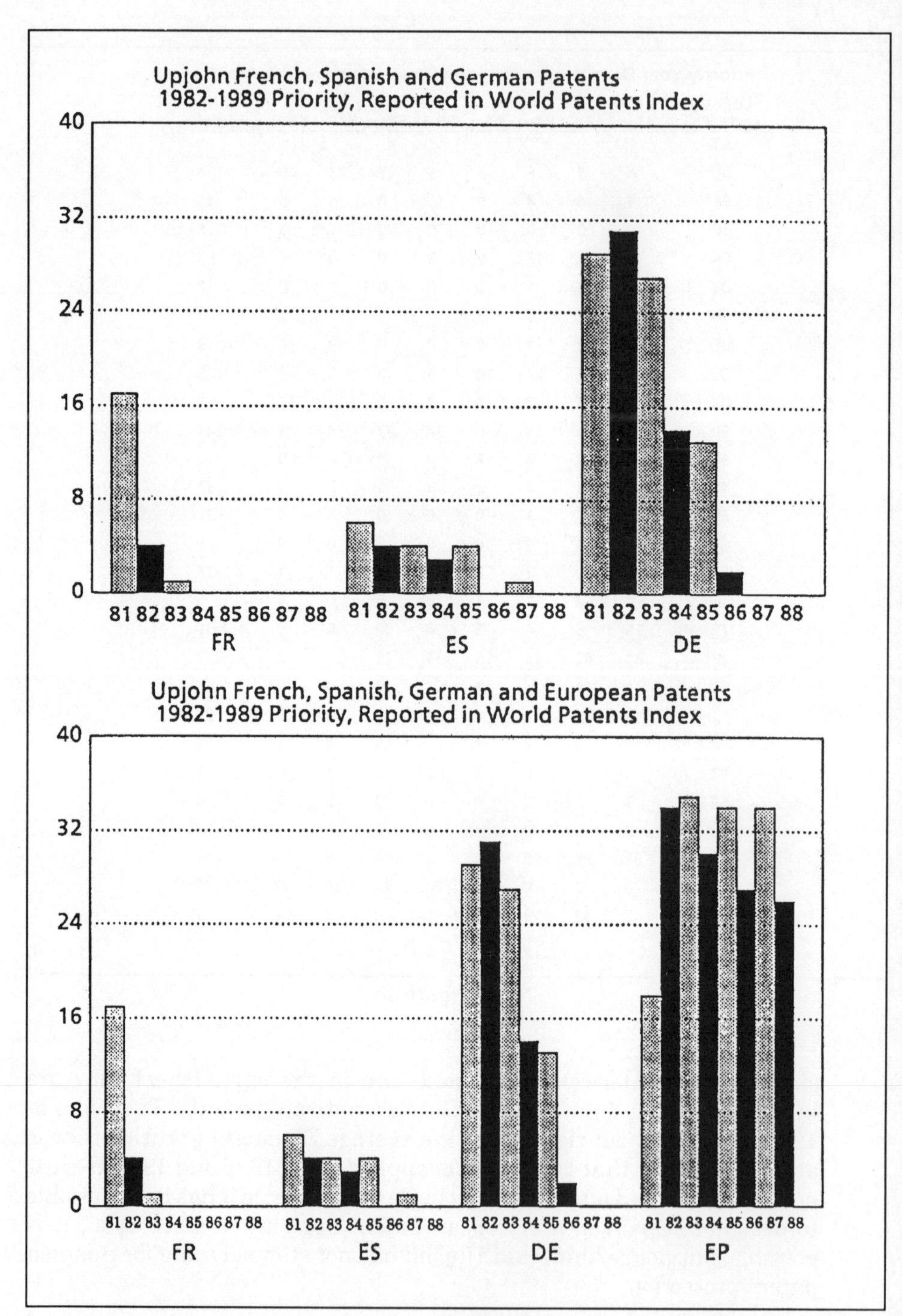
Upjohn French, Spanish and German Patents
1982-1989 Priority, Reported in World Patents Index
40
32
24
16
8
0
81 82 83 84 85 86 87 88
FR
ES
DE
Upjohn French, Spanish, German and European Patents
1982-1989 Priority, Reported in World Patents Index
EP

Figure 15

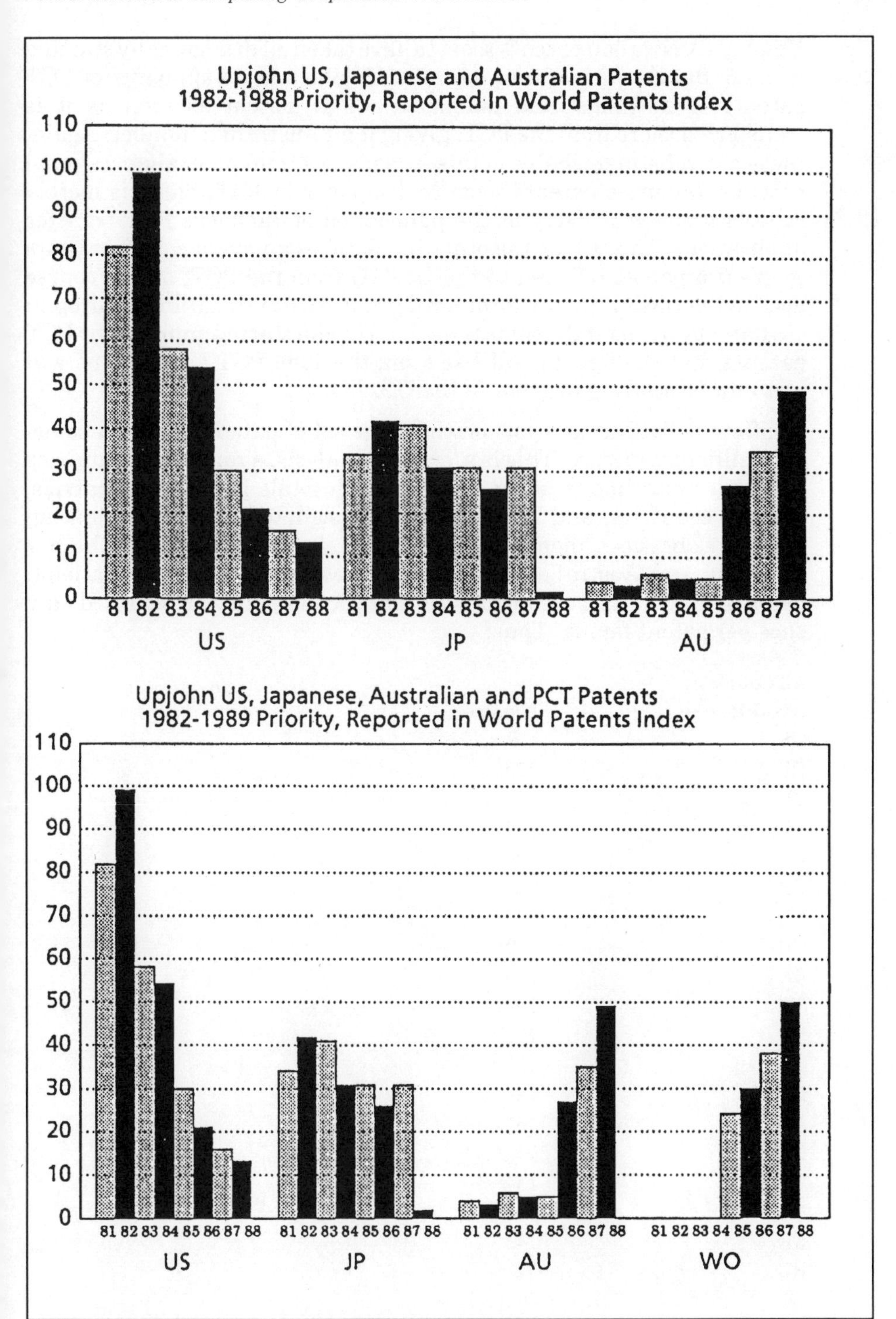
Upjohn US, Japanese and Australian Patents
1982-1988 Priority, Reported In World Patents Index
110
100
90
80
70
60
50
40
30
20
10
0
81 82 83 84 85 86 87 88
US
81 82 83 84 85 86 87 88
JP
81 82 83 84 85 86 87 88
AU
Upjohn US, Japanese, Australian and PCT Patents
1982-1989 Priority, Reported in World Patents Index
110
100
90
80
70
60
50
40
30
20
10
0
81 82 83 84 85 86 87 88
US
81 82 83 84 85 86 87 88
JP
81 82 83 84 85 86 87 88
AU
81 82 83 84 85 86 87 88
WO

Figure 16

Upjohn's Australian patents seem to have taken off dramatically starting in 1986. But this is just one year after Upjohn started applying for PCT patents — and Australia publishes a PCT patent as soon as it is transferred there from the PCT, giving it an Australian number. Again, these are indistinguishable in this spreadsheet from Australian national patents. Japanese patents seem to disappear in 1989, but this merely reflects a one-year delay in the publication of Japanese PCT transfer applications. The US, on the other hand, takes an average of four years to grant a patent referred to the USPTO from the PCT; and of course does not publish it in any form until grant. Hence the dramatic drop in US patents, at about the same time that Upjohn started applying for PCT patents. But the figures will rise soon; the 1986 PCT applications will just now be starting to grant in the US.

A different PatStat analysis on the same set of patents paints a somewhat different picture. This is a 'special' analysis, a ranking of countries in which protection is being sought, which totals patenting countries, priority countries, and EPO and PCT designated states, eliminating duplicates between them. In this analysis, it does not matter whether Germany and Australia showed up as designated states or national patents or both; they are included either way, and they are counted only once per patent family. Thus:

All Countries (PN+PR+DS)	**Number**
US	653
DE	335
JP	333
FR	327
IT	326
GB	322
NL	322
CH	319
BE	309
SE	299
LI	201
AT	191
LU	169
AU	161
DK	158
FI	154
NO	146
KR	143
HU	77
ES	74
CA	71

SU	69
ZA	60
GR	28
BR	14
BG	9
BB	8
CN	8
KP	8
LK	8
MC	8
MG	8
MW	8
OA	8
RO	8
SD	8
PT	7
IL	5
DD	3

The same information appears more dramatically in a bar graph in Figure 17. These data give a far more accurate picture of countries of patenting interest to Upjohn. Unfortunately, this combined- fields analysis cannot produce a country vs. year spreadsheet. Maybe in the next version of PatStat . . . Nevertheless you can see that almost exactly the same number of applications for Japanese, German, and French patent

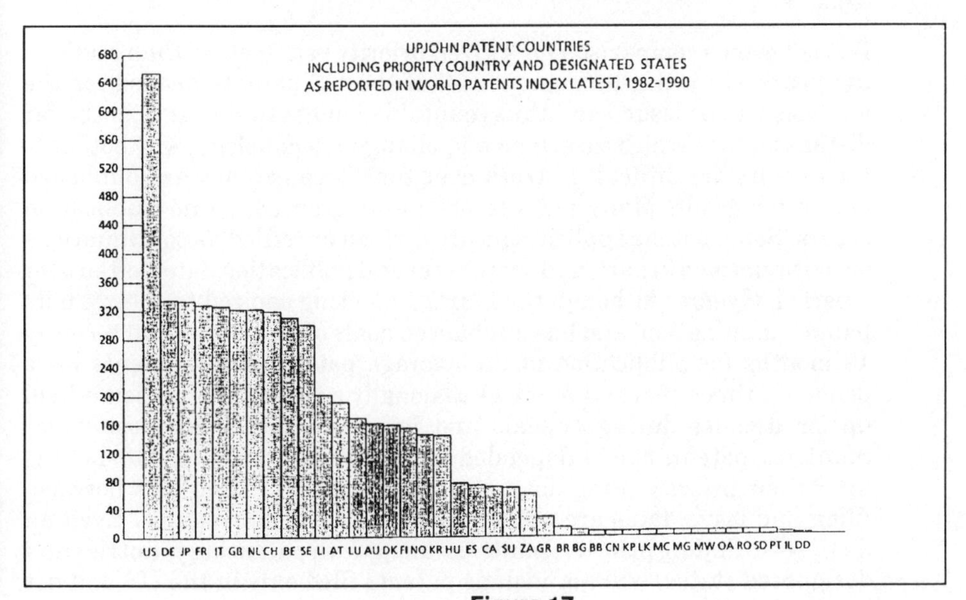

Figure 17

protection have been filed, and that Spain has been designated on some but not all of Upjohn's European patent applications. Furthermore, you can see that Upjohn files a very small percentage of its patent applications in any of the less industrialized PCT member countries. Once again, these inventions could be plums, or maybe only exotic fruit — a PCT application covering a treatment for a tropical disease might well designate Barbados. (Would this patent be a breadfruit?) You cannot distinguish these without reading the patents, and you still may not be able to tell why such broad country coverage is being sought.

Comparisons between the US and other patenting countries

Some patent statistics can demonstrate when other patent statistics will be invalid. For example, if we want to see what countries are considered good markets for a given technology, we can try to determine in which countries patent protection is being sought for that technology. One way to do this is with a patenting-country-plus-designated-states ranking of patents in the technology for a defined time period. However, we must be very careful when we do these patenting country comparisons in fast-growing technologies. If the technology that was a grape a few years ago has grown into a watermelon in recent years — genetic engineering is a good example of such a technology — then a comparison of patenting countries for patents issued/published in any given year will make slow-issuing countries, like the US, look like grapes in comparison with the fast-publishing countries; because US patents issued in a given year reflect priority filing dates a year or more (sometimes much more) older than those of fast-publishing countries' patents published in the same year.

But not even a comparison for a given priority year (rather than publishing year) will work; because not all of the US patents needed for the analysis will be issued and thus countable, unless that year is in the far distant past (in which case, for a fast-changing technology, why bother?). US patents are difficult to track over time because they are published only after grant. Many patents are never granted, so no publication occurs. Some patent applications are divided or refiled as continuations or continuations in part, and acquire several publication dates, often over a period of years. Although the USPTO has long aspired to speed up its patent examination, and has announced goals of 24 months and recently 18 months for publication of the average patent, many patents have pendency times of over 5 years. Occasionally a patent application is held up for decades during appeals and interferences. So in most major countries, patents can be depended on to publish more or less 18 months after their priority filing date; whereas in the US, time lags between filing and issue dates are not only longer, but also irregular. Even an analysis including priority countries (along with patenting countries and designated states) will not include patents filed only in the US and not yet issued. Furthermore, average time lags between filing and issue

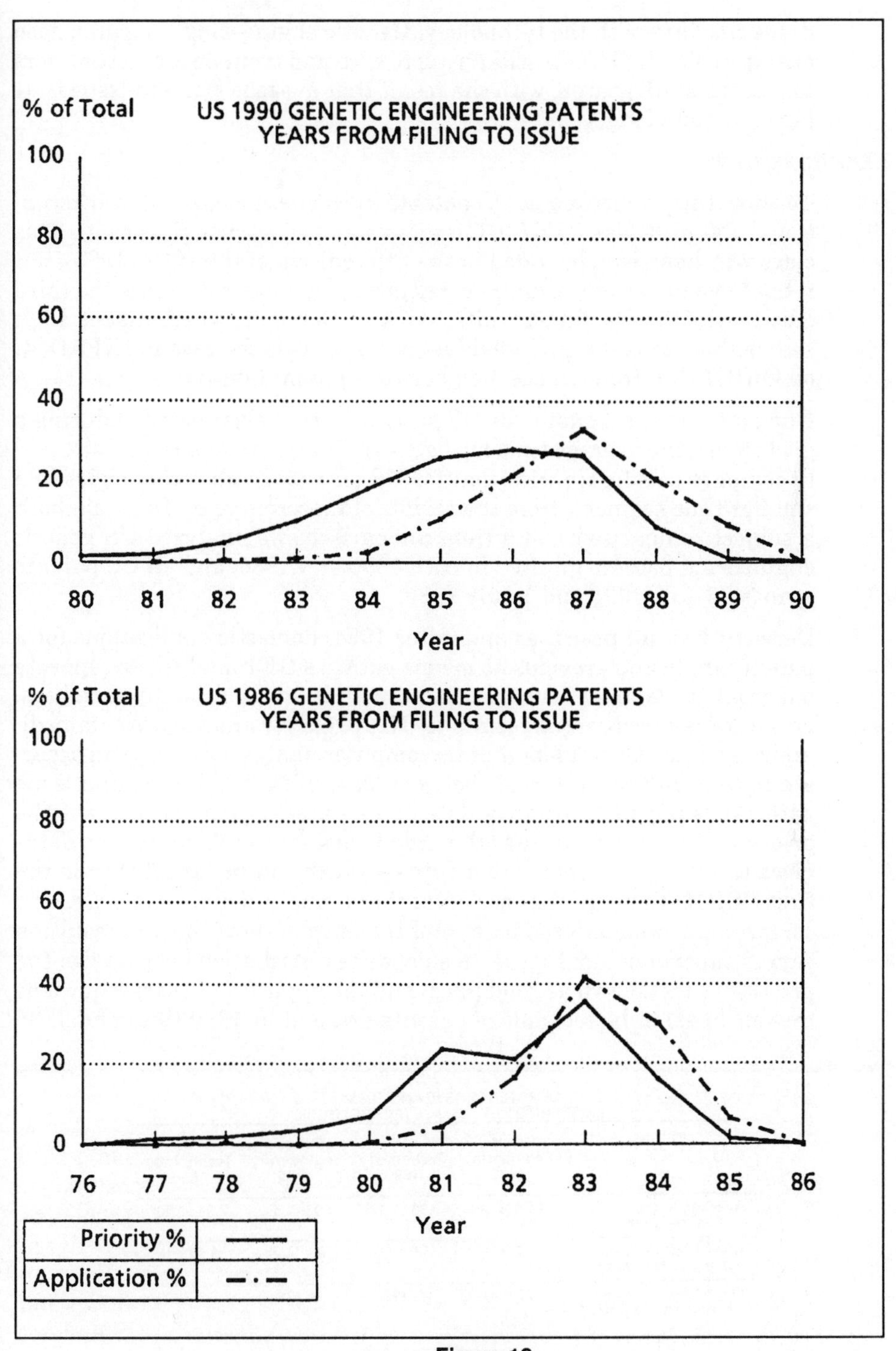
% of Total
US 1990 GENETIC ENGINEERING PATENTS
YEARS FROM FILING TO ISSUE
100
80
60
40
20
0
80
81
82
83
84
85
86
87
88
89
90
Year
% of Total
US 1986 GENETIC ENGINEERING PATENTS
YEARS FROM FILING TO ISSUE
100
80
60
40
20
0
76
77
78
79
80
81
82
83
84
85
86
Year
Priority %
Application %

Figure 18

dates can vary with the technology. Genetic engineering is again a good example. The USPTO is still trying to hire and train enough examiners in this unfamiliar area, with the result that average filing-to-issue time lags are quite long.

Genetic engineering

To show this, we looked at US patents in Derwent classified in International Patent Class C12N-015 (mutation or genetic engineering). This class was heavily subdivided in the fifth edition of the IPCs (1989); but it has been in existence, and covered genetic engineering, since the third edition (1979). We did our initial look in Derwent, which makes IPCs searchable not just from individual patents (as is the case in INPADOC on ORBIT) but from all the members of a patent family.

Our aim was to calculate, for US patents in that class granted during a given year, the average time lag between filing date and issue date, and to compare this figure with that for US patents in all technologies in a smaller time segment, from the middle of the same year. To obtain both a subject comparison and a time comparison, we analyzed US genetic engineering patents granted in both 1990 and 1986, and all US patents granted 3 July 1990 and 1 July 1986.

Derwent lists all priorities and (after 1984) domestic applications for a patent family and provides no means within a GET analysis to separate out priorities/applications associated with individual family members. So we transferred search results to INPADOC for analysis. We immediately ran into the problem that no computer analysis would yield accurate figures on the time lapse between date of first US filing and issue date. An analysis of priority dates give dates a year too early if the priority was non-US. On the other hand, an analysis of application dates gives the latest US application date — which can be far later than the first USPTO filing, in the case of continuations and divisionals. So, as a compromise, we analyzed both; and the graphs show the two resulting curves superimposed. Figure 18 shows the distribution over time of the priority years (shown as solid lines) and the application years (shown as broken lines) of biotechnology patents granted in 1990 (top) and 1986

US PATENTS, 1990 AND 1986:
BIOTECHNOLOGY VS. ALL TECHNOLOGIES

	Genetic Engineering 1990	All Technology 1990	Genetic Engineering 1986	All Technology 1986
Average filing to issue time, years	4.74 (3.16)	2.93 (1.77)	3.93 (2.51)	2.91 (2.00)
Time till 95% of patents are granted, years	7.7 (4.9)	6.0 (2.9)	6.4 (4.2)	5.3 (3.1)
Percentage of patents issued after 3 years	28 (64)	77 (95)	48 (78)	75 (95)

Figure 19

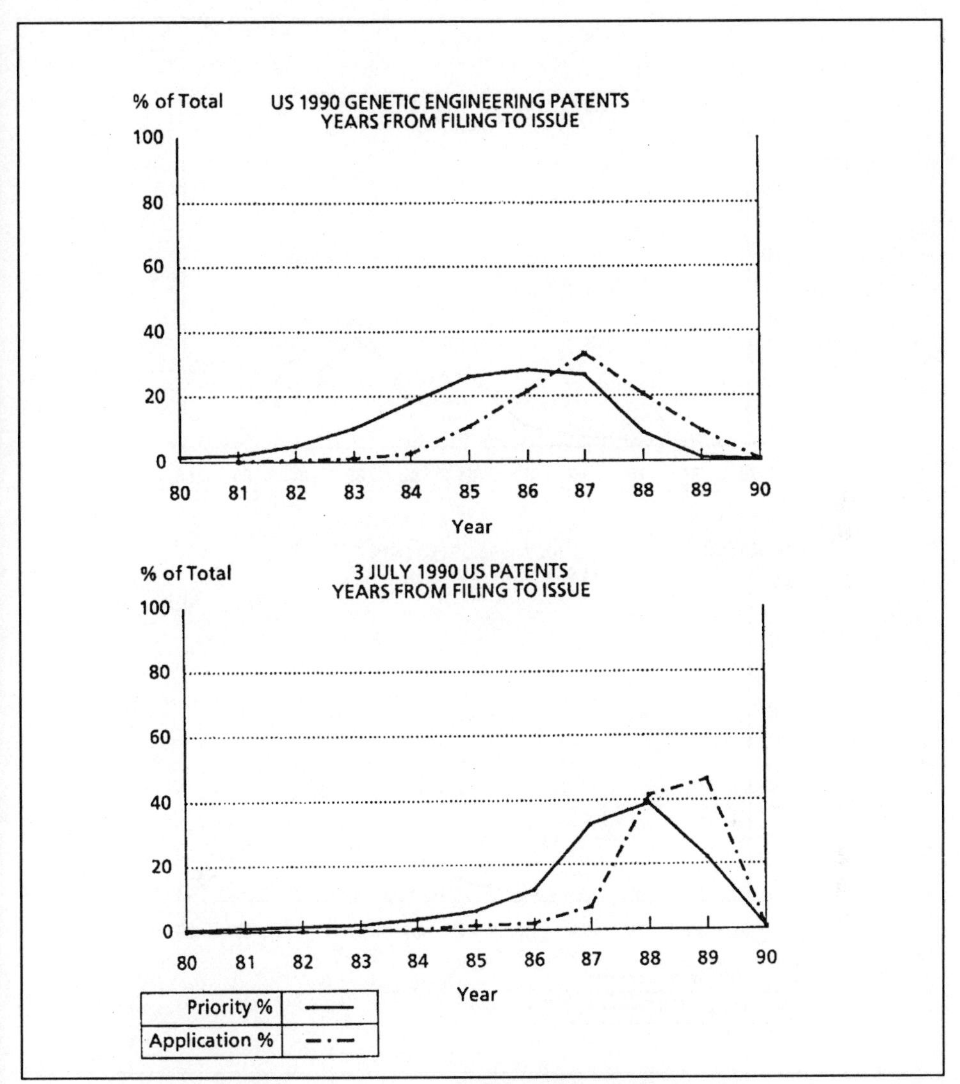

Figure 20

(bottom). The figures tabulated in Figure 19 and summarized below will give results from the priority year analyses first, followed by results from the application year analyses in parentheses.

Calculations showed that the 1990 genetic engineering patents had an average pending time in the USPTO of 4.7 (3.2) years, compared to 2.9 (1.8) years for US patents in general. Even worse, the time needed for 95% of the genetic engineering patents to have issued (thus permitting some sort of significant statistical calculations) was 7.7 (4.9) years, compared with 6.0 (2.9) years for US patents in general. Only 28% (64%)

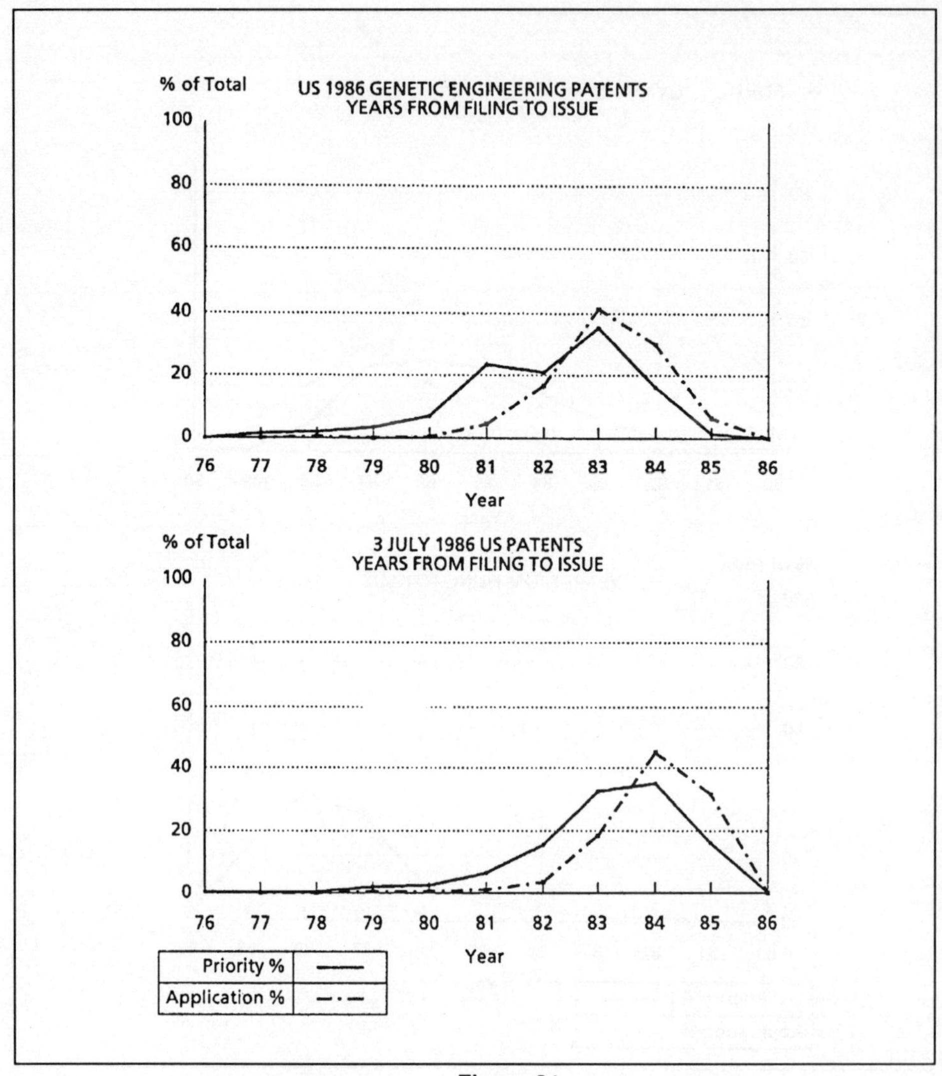

Figure 21

of the genetic engineering patents had a pending time of 3 years or less, compared with 77% (95%) of US patents in general. This contrast is shown graphically in Figure 20.

And the time lags for genetic engineering are getting worse, while those for US patents in general are improving somewhat. The analysis of 1986 patents, shown graphically in Figure 21, showed an average pending time in the USPTO for genetic engineering patents of 3.9 (2.5) years, compared with 2.9 (2.0) years for US patents in general. 95% of the genetic engineering patents had issued after 6.4 (4.2) years, compared with 5.3 (3.1) years for US patents in general. 48% (78%) of genetic engineering patents had a pending time of 3 years or less, compared with 75% (95%) of US patents in general.

We conclude that we cannot perform the country distribution we wanted in this fast-changing, slow-issuing technology. Three years from their US filing dates, by which time virtually all of their foreign counterparts will have been published, a very significant — and incalculable — number of US patents in genetic engineering will not yet have issued. So comparing country coverage in this technology is indeed a grapes-and-watermelons situation.

Conclusion

So, gather your patent statistics; juggle your apples, oranges, grapes, and watermelons; enjoy your plums. But do not ever assume that you are working with a homogeneous bowl of mashed bananas.

Acknowledgements

We would like to thank Richard Kurt and Maxwell Online Inc. for their extremely generous donation of ORBIT online time; Andrea Rosanoff for her statistical calculations; Stuart M. Kaback for valuable suggestions and input; and our respective pets (including Edlyn's husband Bill) for their patient tolerance of our intolerable absences from home during the preparation of this paper.

References and Notes

1. A great many papers have been published on the statistical analysis of patents. These are only a few:

a) Mark P. Carpenter, Francis Narin and Patricia Woolf. 'Citation Rates to Technologically Important Patents.' *World Patent Information* **3**(4), 160-163 (1981).

b) L. O. Levine. 'Patent Activity by Date of Application — Estimating Recent Applications in the U.S. Patent System.' *World Patent Information* **9**(3), 137-139 (1987).

c) Francis Narin and J. David Frame. 'The Growth of Japanese Science and Technology.' *Science* **245**, 600-605 (11 August 1989).

d) Mary Ellen Mogee. 'International Patent Data for Technology Analysis and Planning.' Mogee Research and Analysis Associates, 212 Carrwood Road, Great Falls, VA 22066 (October 1990).

2. a) H. Aspden. 'Patent Statistics as a Measure of Technological Vitality.' *World Patent Information* **5**(3), 170-173 (1983).

b) Stuart M. Kaback. 'Online Patent Information — Patent statistics and other games.' *World Patent Information* **6**(2), 80-81 (1984).

c) R. M. Wilson. 'Patent Analysis using Online Databases — I. Technological Trend Analysis.' *World Patent Information* **9**(1), 18-26 (1987).

d) R. M. Wilson. 'Patent Analysis using Online Databases — II. Competitor Activity Monitoring.' *World Patent Information* **9**(2), 73-78 (1987).

e) Edlyn S. Simmons and Stuart M. Kaback. 'Online Patent Information — Doubleheader.' *World Patent Information* **10**(3). 204-6 (1988).

f) J. van der Drift. 'Statistics of European Patents on Legal Status and Granting Data.' *World Patent Information* **10**(4), 243-49 (1988).

g) M. Hill. 'Information from patents — an overview of recent activities.' *Online Review* **13**(3), 199-204 (1989).

h) Nancy Lambert. ' Patent statistics searching: practices and pitfalls.' In Proceedings of the Montreux International Chemical Information Conference & Exhibition, H. Collier, Ed., Springer Verlag: Heidelberg, 1989, pp 141-160.

3. By T-test, alpha level 0.05.

4. By approximate T-test, $p < 0.01$.

Source of cited references	Reasons for citation	Availability of information about cited references
Examiner's search	Describes invention	Published in patent document or search report by some patent offices
	Renders claimed invention obvious	Indexed in some databases
	Shows state of the art	
Applicant's citations in patent specification	Describes background of the invention	Published in patent specification
	Summarises unsuccessful alternatives to claimed invention	Some references included in examiner's citations
	Describes methodology for making, using or testing claimed invention	Others searchable only in full-text databases
Cited by applicant in correspondence or information disclosure statement	Describes similar but distinct inventions	Not published
	Describes unsuccessful alternatives to claimed invention	Some references included in examiner's citations
	Lists references cited by other patent offices	Some patent office records can be obtained for review

Addendum: Figure 11a: Sources of Patent Citation Data

Patenting in biotechnology: An overview of current issues

R.S. Crespi

Week by week the world's store of patent information increases through the official publication of granted patents and pending patent applications in all fields of technology. US patent disclosures are not published until the patent is granted whereas in Europe, Japan and some other countries the patent disclosure is laid open at the application stage at which no official examination of the prospective rights of the applicant has begun. Under the US system the final claims of the patent have emerged after official scrutiny, which is normally rigorous, whereas under the early publication system of European patent law the claims in the published document represent the scope of protection the applicant thinks himself entitled to, or would like to have if he can persuade the Examiner to accept them.

The early publication system originated as a measure designed to avoid the long delays in making patent information available to the public due to the official backlog in the processing of applications to grant. It commanded wide assent throughout industry, which was apparently prepared to trade off the consequences of each disclosing his own hand at the beginning of the game in return for seeing the hands of all his competitors at a similar stage. The information industry is likewise grateful for this bounty.

Official spokesmen in the field of patent information can produce statistics to show that the patent literature is a valuable resource which must not be overlooked because a high proportion of these disclosures are not published in the scientific literature. This is less true in the field of biotechnology than in others. Many biotechnology inventors are based in academic research and follow the scientific tradition in this respect. These may be funded by private or public sponsors for whom the insistence on scientific publication is acceptable provided the foundation of patent protection has first been laid.

Problems of Biotechnology patenting

The special problems of biotechnology patents began to emerge with the development, after the second world war, of microbiological processes for producing antibiotics, amino-acids, enzymes and other microbial products of industrial importance. To approach these problems it was inevitable that inventors and patent experts would draw upon a century or more of experience in the patenting of chemical inventions and apply the established principles to the new situations. On the whole this has been successful, due to the fair degree of parallelism that exists between the two technologies. But the significant differences between living cells and the more manageable inanimate molecules have brought about a re-examination of traditional ideas of patent law to adapt it to deal effectively with biological systems and so stimulate innovation in this new field as it has done for older technologies.

When a new chemical process or product has been devised, the conditions and the steps necessary to obtain patent protection are relatively straightforward. In addition to being novel the process or product must be inventive i.e. it must be more than just an obvious progression from what is already known, and it must have industrial applicability or some other form of practical utility. The inventor must also provide a description which the ordinary skilled chemist can follow in order to repeat the process or prepare the product — in patent parlance this is known as an "enabling disclosure".

For example, to provide an enabling disclosure of a chemical synthesis starting from known compounds, the methods can usually be described in sufficient detail to enable the products to be prepared in the reported yields. It will be normal practice to include examples of the preparative methods containing experimental details such as would be required for a paper in a scientific journal. This disclosure requirement applies primarily to the patent application but it is also relevant to the question of novelty over prior art. Thus a patent application cannot be rejected as lacking novelty over a prior publication unless the publication discloses sufficient information how to make the compounds, at least when taken in conjunction with what the ordinary chemist is assumed to know.

Microbiological inventions follow the chemical pattern to a large extent and fit into the following main categories.

Production of micro-organism	(Process)
New Micro-organism	(Product)
End Products of biosynthesis	(Product)
Formulations of micro-organism	(Composition)
Use of micro-organism	(Use/process)

to produce e.g. . . .

- biomass
- by-product of microbial growth
- an extracted product
- an improved substrate
- a biotransformation product.

The microbiological equivalents of chemical products will be microbially produced products, or new micro-organisms themselves or some other type of biological material including cell lines and plant or animal cells.

The Enabling Description of Micro-organisms

It is a fundamental requirement of patent law that in return for legal protection, an inventor must disclose the invention in a manner sufficiently clear and complete to enable others of ordinary skill in the art to repeat or reproduce the process or product for which the patent is granted. Where the invention consists of or depends on a specific micro-organism or other kind of biological material, this must be identified in the patent application to provide the 'enabling disclosure' the law requires.

As compared with the chemical case, where a written description of the process is usually adequate, the requirement to provide a description which is reliably repeatable presents greater difficulty when living material is involved. Living material does not admit of a total description. But, more importantly, a total description would not by itself guarantee the means of producing the material. Where the micro-organism is known and already available to the skilled person and the invention resides, for example, in the discovery of some new property or use of practical value, it is usually sufficient to refer to the micro-organism by name. But for a new micro-organism, say a newly isolated or developed strain of a known species, the skilled person who attempts to repeat the procedure described in the patent specification will in most cases need not only a description of the organism but also a means of access to it.

Now if the patent application gives reliable instructions how to re-isolate, re-discover, or reconstruct the new organism this will be a sufficient disclosure. However, in most cases this cannot be achieved with certainty. Patent law has solved this problem by making use of the Culture Collection deposit system which the scientific community had created much earlier for its own needs.

Depositing micro-organisms in Culture Collections for the purposes of patent applications is now widely established. The practice has developed internationally both through case law and in the express obligations written into modern patent laws in many countries e.g. the Budapest Treaty. The maxim that what cannot be fully described must be deposited has therefore become part of patent law. This development has added a new dimension to patent law and practice for which no parallel exists in chemistry or other fields of technological innovation.

Thus the deposit of the organism supplements the written description and fulfils other important functions. First, it provides a reference material for resolving any dispute over the alleged novelty of the organism. Secondly, its reference function may be called upon to decide whether any third party is infringing the patent by using the same organism without a licence from the patentee. Finally, the deposit provides an available source material to enable others to make use of it when they are legally free to do so ie, when the patent is allowed to lapse or expires at the end of its normal term.

Availability of the deposit

As to when a deposited culture can become available to the Applicant's competitors and other third parties there is a difference between both the US and Japanese patent systems, which allow access to the deposited culture only after an enforceable right has been granted, and the corresponding laws in European countries which allow access to the deposited culture upon first publication of the European or National Patent application. The drawbacks of the European law on this topic have been emphasised by industry from the very beginnings of the European Patent Convention itself and the efforts of interested circles to improve the law have continued unabated since that time. A comparison of the US, European, and other national rules on the release of patent cultures is shown in the accompanying Figure.

Loss of control of the new strain, at least for competitive research purposes, is mitigated under European patent practice by the option to elect for the so-called independent expert solution in the interim period between publication of the application and eventual grant of the patent. The function of the independent expert is to act for other persons or firms, eg the applicant's competitors, and to test, experiment with, and generally evaluate the invention on their behalf. The independent expert, though acting for third parties, must not pass the strain on to them. Under this alternative a measure of control on the use of the organism is provided for the applicant before he obtains an enforceable right.

Eventually there must be full disclosure to the public and the skilled person must be put in a position of reproducing the inventive process or product, including access to deposited biological material. From then on policing problems vary with the nature of the patent protection obtained. When the microbiological invention leads to new products which can be the subject of product patents then detection of infringement is relatively straightforward. Where the products are not new, however, and novelty lies only in the strain used or in some other parameter of the new microbiological process then policing becomes more difficult. These factors can usually be fully anticipated before the decision to proceed by the patent route is first taken. One crumb of comfort may lie in the fact that the laws of some countries allow infringement actions to proceed on the basis of a strong *prima facie* case that the defendant is using the patented process. If the record shows that the defendant has obtained a sample of the patentee's deposited strain from the culture collection this could well be sufficient to shift the burden of proof away from the patentee.

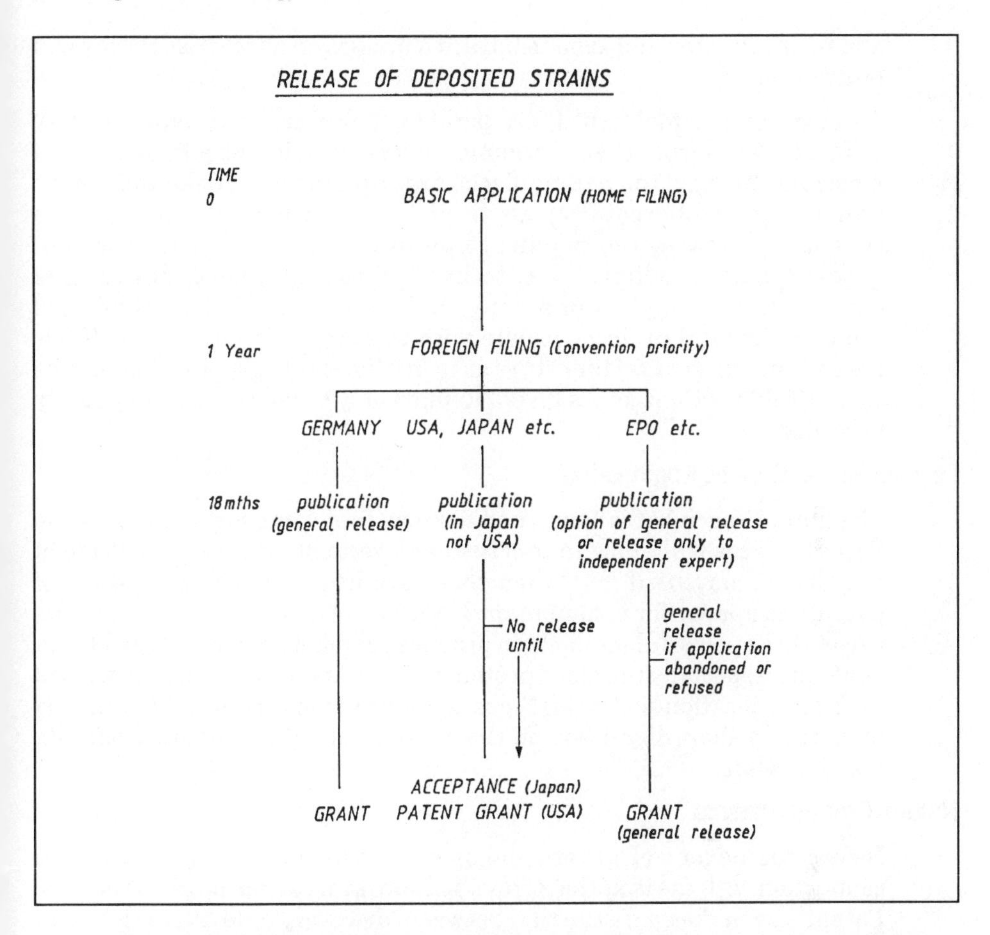

The compulsory release of deposited cultures at the early publication stage without geographical limitation is the greatest single disincentive to the use of the patent system for the protection of microbiological inventions. The final solution of this difficulty would be to substitute the date of grant for the date of publication of the European application. However this has been dismissed by official circles as impossible because the publication without the culture would not constitute an enabling disclosure and would therefore not be usable by them as prior art against later applicants.

Genetic engineering patents

In the field of classical microbiology, where the prototype patent specification describes the production of an antibiotic or other product of microbial culture, the experimental and other parts of the disclosure follow a pattern which is familiar from our experience with chemical patents. For these it is therefore not difficult to appreciate the novelty and inventive step and even to hazard a prediction as to what claims will be granted.

To come to recombinant DNA patents, however, is to enter a very different world from that of organic chemistry. Without a knowledge of molecular biology, the patent disclosure will probably strike the patent searcher as so different from anything he has been accustomed to as to constitute a new species of patent literature. There will be a statement of object, a list of definitions of technical terms, flow sheet drawings of cloning strategy, and an impressive but densely detailed description of the experimental protocol replete with so many references to methods taken from the general literature as to justify a bibliography. Sometimes even a Table of Contents is given. Remarkably broad claims will usually be added.

Patentability in Genetic Engineering

The patent specialist is concerned to know first what sort of patents the Patent Offices will allow in this field and secondly what the attitude of the Courts may be if and when these are litigated. The first wave of patents to appear for recombinant DNA inventions were directed to the use of these powerful methods to produce proteins known to be useful in medicine, especially the blood proteins. These are conveniently described as first-generation products. Since these proteins are already produced in Nature a short digression on the question of natural product patents may be helpful.

Natural Product Patents

Before the advent of genetic engineering this question first arose in connection with the isolation of purified substances from natural sources. Patent law makes a distinction between 'discovery' and 'invention' and European law specifically excludes mere discovery from patentability. Some might say that the isolation of substances from Nature is mere discovery and not invention. The European Patent Office has rejected this view, at least where the product is "new" in the absolute sense of having no previously recognised existence, in which case the substance per se may be patented.

The patenting of purified natural products has been extensively covered in the case law of the United States. The early court decisions pointed to the conclusion that a known substance cannot be patented as a pure product solely on the basis of its purity. However, where the substance was known previously only in the form of crude extracts which were of no use to man or beast, an inventor who had devised a practically useful method of making pure product in quantity could patent not only the specific method but could claim the pure product per se. The classical example of this was the production of Vitamin B_{12} by fermentation and its isolation in bulk as the pure substance. Previously the substance had been available only as crude liver extracts which were of no use for therapeutic application. A US District Court [1] upheld a product per se claim to the pure vitamin defined by chemical and physical properties.

At present, this basic question is being considered once again in current US litigation over patents for purified naturally occurring proteins including Factor VIII, the blood clotting factor important in the treatment of hemophilia, and Erythropoietin, the hormone which stimulates the formation of red blood cells. In both cases the patent discloses a particular method of purifying or isolating the protein. However, the claims are not limited to products produced by these methods but also cover the proteins defined solely in terms of units of activity, a parameter connected with the degree of purity which these methods have made possible. The following is a short account of the way the US courts have approached these issues.

Scripps Clinic and Research Foundation v Genentech [2]

Scripps US patent 4,361,509 covered a method of separating Factor VIII procoagulant activity protein (VIII:C) from the von Willebrand factor protein (VIII:RP) with which it is complexed in plasma. The method of preparation of the purified VIII:C component described in the patent involves removal of VIII:RP from the VIII:C/VIII:RP complex present in plasma or a commercial concentrate by adsorption on a specific monoclonal antibody.

The patent has claims to this particular method of purification and claims to the product of such a method.

Scripps realised later that it was desirable to have product claims to the purified product broader than those they had obtained in the patent. They therefore filed for an additional patent and obtained Reissue patent 32,011 which has the folowing product claim:—

> *A human VIII:C preparation having a potency in the range of 134 to 1172 units per ml and being substantially free of VIII:RP*

This claim is directed to a product in terms of its potency and purity but makes no mention of the method by which it is obtained. It would therefore appear to cover a purified VIII:C made by any method whatsoever. It would thus cover Factor VIII:C produced by recombinant methods and having the specified potency. Genentech had produced the recombinant product and were sued by Scripps for infingement of their reissue patent. Genentech had themselves applied to patent their technology but even if they were to succeed it would not affect the question whether their product is dominated by the Scripps patent. In the meantime a patent on a recombinant method for producing VIII:C had been issued to Genetics Institute Inc.

The Scripps-Genentech litigation is now moving towards a conclusion. A similar legal point is involved in the following case:—

Amgen Inc. v Genetics Institute Inc. [3]

Erythropoietin is the natural protein which stimulates the production of red blood cells. Genetics Institute Inc. had filed an application on a method of purifying erythropoietin from human urine. In the resulting GI patent, US 4,677,195, the GI method of purification was by reverse phase high performance liquid chromatography but, unlike in the Scripps case, GI made sure of having broad product claims in this patent, including the following:—

> *Homogeneous erythropoietin characterised by a molecular weight of about 34,000 daltons on SDS PAGE, movement as a single peak on reverse phase high performance liquid chromatography and a specific activity of at least 160,000 IU per absorbance unit at 280 nanometers.*

Amgen holds US patent 4,703,008 for the preparation of erythropoietin by recombinant methods. However, the Amgen patent claims DNA sequences coding for the protein as well as the corresponding vectors and transformed cells, but the recombinant protein is not claimed as such. The prototype of the Amgen claims is the following:—

> *A purified and isolated DNA sequence consisting essentially of a DNA sequence encoding human erythropoietin*

The two instances outlined above are examples of conflict between a natural product patent and a recombinant DNA patent in which it has to be decided who is free to operate commercially with the recombinant protein. In these two examples, especially the second, one question for the court is whether a claim based entirely on the purification of the natural product can be permitted to cover the protein as made by recombinant methods. But the more basic issue is whether claims to purified natural products are legally valid in broad terms such as those used in the above examples.

The New biotechnology (Recombinant DNA)

In a recombinant DNA patent, although there are some differences from one country to another, it will be common to have claims of the following types:

a) Recombinant protein products (and alleles, variants, derivatives)

b) DNA sequences coding for the products of (a)

c) Vectors containing the DNA sequences of (b)

d) Micro-organisms, cell lines, and other organisms transformed with vectors (c)

e) Processes for constructing the micro-organisms etc. of (d)

f) Processes of producing products as in (a) by expression of DNA sequences (b) in a recombinant host organism

The Genentech v Wellcome case [4] on tissue plasminogen activator, under Genentech's UK patent 2,119,804, is a prime example of reliance on a claim of type (a) viz.

Human tissue plasminogen activator as produced by recombinant DNA technology

On commercial grounds this type of claim was exactly what was wanted but it failed on legal grounds, being held obvious, first, because a product of this kind was a known desirable objective and, secondly, because the product was obtained by using known methods of gene cloning. Thus, for the British Courts, being first in the race to clone and express genes for known proteins is not enough to support so broad a legal monopoly in such circumstances. The main problem here for the patentee was the absence of any commercially valuable fall-back position in the other claims of the patent which the court might have been prepared to sustain. There is no example yet of a non-British court taking the same line as in the t-PA case.

Had Genentech been the first to produce t-PA as a pure therapeutically usable product, the complexion of the case would have been quite different. A patent for a new second generation derivative or analogue of t-PA e.g. obtained by mutagenesis of the natural DNA and showing improved properties would be treated more leniently by the British court.

Hybridoma Technology

The patenting of hybridomas and other cell lines is based in many countries on the principles established for micro-organism patents. Patents can be granted for novel parent myelomas used to make hybridomas, the family of derived hydridomas and specific hybridomas, the production of monoclonal antibodies in process or product terms, as well as applications of these in therapy or diagnostics.

Many patents are being granted based on the use of monoclonal antibodies and systems containing them in diagnostic assays. In these the claims are directed either to the method of assay or to particular reagent compositions or to combinations of materials useful in diagnostic kits.

After the publication in 1975 of the basic Milstein/Kohler technique and the subsequent appreciation of its more general importance it was reasonable to assume that the patentability of any application of this general procedure must rest on some special and non-obvious property or advantage of the particular system constructed.

In the Wistar Institute case [5] the following claim came under scrutiny in the British Patent Office:

> *A process for producing viral antibodies comprising fusing a viral antibody producing cell and a myeloma cell to provide a fused cell hybrid, culturing said hybrid and collecting the viral antibodies.*

The only distinction between this and the prior known process appears to rest on the term "viral".

The UK Patent Office took the view that since the basic monoclonal antibody technique had already been published it was obvious to apply this technique to an area where it had already been considered valuable, ie, for viruses. The Patent Office also would not allow a specific claim to the particular hybridoma developed for producing monoclonal antibodies against influenza virus. The applicant had not shown that the preparation of hybridomas of this type required anything other than the application of known techniques and moreover none of the hybridomas could be said to have particularly unusual advantages. It must be pointed out however that the US Patent office did not take the same approach and in fact granted US patent 4,196,265 containing this broad process claim.

In the Hybritech case (US) [6] the patentee Hybritech successfully sued a number of companies under its US patent 4,376,110. The invention lay in the use of monoclonal antibodies in place of prior art polyclonal antibodies in a known sandwich assay system. When the US Patent Office originally examined the application, it argued that it would be obvious to use monoclonal antibodies in place of polyclonals in conventional immunoassay protocols. This objection was overcome by including in the claims a numerical limitation regarding the affinity (binding power) of the antibodies. In the District Court the patent was held invalid on the grounds of obviousness but this was over-ruled by the Court of Appeals for the Federal Circuit because the prior art was "devoid of any suggestion that monoclonal antibodies can be used in the same fashion as polyclonals". Also it was influenced by the commercial success of the patentee's product and found that a 3 year time gap between the first availability of monoclonals and the sale of the patentee's kits was long enough to indicate lack of obviousness.

By contrast, in Unilever PLC v Boehringer Mannheim [7] (immunoglobulins) the Technical Board of Appeals of the European Patent Office has held that the replacement of polyclonal antibodies by monoclonal antibodies in an immunopurification process required no inventive step at the priority date of the opposed patent (February 1981).

The Unilever claims were to a process for the recovery of immunoglobulins of high purity and potency from milk by selective binding to a low affinity monoclonal antibody specific to one or more of the immunoglobulins and itself bound to an insoluble carrier material. It was clear that the novelty lay only in the use of monoclonal antibodies in substitution for the prior use of monospecific polyclonal antibodies in the same basic process. This substitution was held to be the "next logical step" and the "desired logical and obvious step" to improve the purification process of the nearest prior art.

On the other hand, where the invention has been assessed by the European Appeal Boards as in the 'pioneering' category the patents have been upheld with extremely broad claims. Two examples are as follows:

Polypeptide expression (Genentech I) [8]

In EPC application 001,929 corresponding to Genentech 'cloning vector' US patent 4,707,362, the invention was covered in the main claim directed to a recombinant plasmid the most important features of which were the presence of a homologous regulon and a heterologous DNA coding for a desired polypeptide and in proper reading frame. In this context homology is to be understood in relation to the host micro-organism intended for transformation with the claimed plasmid.

In spite of close prior art showing the use of the claimed homologous/heterologous combination to produce a transcript which would not have been translated into protein, coupled with a speculation as to the possibility of expressing eukaryotic DNA in bacteria, the Board found in favour of the applicant on the issue of inventive step, since the prior document was judged as pointing away from the invention.

This was one of the earliest applications in the field of genetic engineering, and the Examining Division had objected to the use, in the claims, of the broad functional terms 'bacteria, regulon, plasmid etc', arguing that these should be limited to specified known materials in order to meet the requirement of sufficiency and to avoid covering future inventions in the discovery of new materials falling within these broad descriptions.

The Board of Appeals followed its prior ruling in a chemical case allowing functional terms in appropriate situations, and held that this must apply also to the field of biotechnology. The terms used in this application could generically embrace the use of unknown or not yet envisaged possibilities, including specific variants which might be provided or invented in the future. For the invention described here only the use of such functional terminology could give fair protection.

Conclusions

It will be apparent that biotechnology patent issues are highly controversial not only as between industry and official patent authorities but also within industry itself, especially among the biotechnology companies seeking to secure dominance over selected areas of the technology which have enormous market potential. The major problem attaches to the evaluation of inventiveness as regards broad claims for recombinant products and for the DNA molecules which code for their biosynthesis. Consistency of approach by the courts and patent offices of different countries has not yet been reached, a fact which renders it more than usually difficult to make the predictions that patent lawyers are frequently asked to provide.

From the few examples given above it is still early to predict how the legal pattern will eventually be drawn. The climate for obtaining and defending US patents in this area certainly seems to be favourable. The European Patent Office also seems anxious to sustain biotechnology patents, at least at the Appeal Board level.

The British t-PA decision, though apparently against the trend in the USA and Europe, supports the view that merely to be first in the race is not enough to stop all others from running along parallel paths that they have devised from the known and available technology. The British t-PA decision, it must be stressed, applies only to first generation products. As regards next generation products of the t-PA type, numerous patent applications are now in the pipeline for modified or mutant forms of natural t-PA. These being new compounds, their patentability can be assessed according to well-known criteria. It is clear that patent prospects are reasonable for these compounds.

References

UK case citations are to 'Reports of Patent Cases' (RPC). US case citations are to United States Patent Quarterly (USPQ). EPO decisions are cited as case T numbers.

[1] Merck & Co v Chase Chemical Co., 155 USPQ 155

[2] Scripps Clinic & Research Foundation v Genentec, 11 USPQ 2d 1187

[3] Amgen Inc v Genetics Institute/ Chugai, 10 USPQ 2d 1906, 1909 (1989) and 13 USPQ 2d 1737-1798

[4] Genentech v Wellcome Foundation, 1989 RPC 1473. See also Crespi, R S — 'Claims on tissue plasminogen activator' *Nature* Vol **337**, 26 January 1989. 317.

[5] Wistar Institute's application 1983 RPC 255

[6] Hybritech v Monoclonal Antibodies Inc., 231 USPQ 81

[7] Unilever v Boehringer Mannheim, Technical Appeal Board Decision T 499/88

[8] Polypeptide Expression (Genentech 1), Technical Appeal Board Decision T 292/85

Searching techniques for the tertiary structures of proteins in the Protein Data Bank

Peter J. Artymiuk, Helen M. Grindley, David W. Rice, Elizabeth C. Ujah and Peter Willett[1]

Krebs Institute for Biomolecular Research, Departments of Information Studies and Molecular Biology and Biotechnology, University of Sheffield

Abstract

This paper describes work to date on a project to develop graph-theoretic techniques for the representation and searching of the 3-D protein structures in the Protein Data Bank. The proteins are represented by labelled graphs, in which the nodes of a graph describe the α-helix and β-strand secondary structure elements and the edges of a graph describe the geometric relationships between pairs of these elements. Two programs are described, POSSUM and PROTE, that allow the searching of such graphs using subgraph isomorphism and maximal common subgraph isomorphism algorithms, respectively.

1 Introduction

The last three decades have seen the development of a range of sophisticated computer-based systems for the storage and retrieval of information pertaining to chemical structures [4,17]. A molecule in such a system is generally represented by a *connection table*, which contains a list of all of the non-hydrogen atoms, together with bond information that describes the exact manner in which the individual atoms are linked together.

---

[1] To whom all correspondence should be addressed at: Department of Information Studies, University of Sheffield, Western Bank, Sheffield S10 2TN, England

An important characteristic of a connection table is that it can be regarded as a *graph*, a mathematical construct that describes a set of objects, called *nodes* or *vertices*, and the relationships, called *edges* or *arcs,* that exist between pairs of the objects. A connection table is an example of a *labelled graph*, since the nodes and edges have labels associated with them, i.e., the elemental types and the bond types, respectively. This equivalence means that searching operations on databases of connection tables can be implemented using *isomorphism* algorithms, which compare one graph with another to determine the structural relationships that exist between them. Specifically, chemical searching and substructure searching correspond to *graph isomorphism*, the comparison of one graph with another to determine whether they are identical, and to *subgraph isomorphism*, the identification of the presence of a query graph within a larger graph, respectively. In addition, *maximal common subgraph isomorphism* algorithms are used for the detection of the reaction sites in chemical reaction database systems and for structure elucidation studies, *inter alia*.

Until recently, the use of graph-theoretic algorithms in chemical information science has been restricted to the processing of 2-D molecules. The techniques that have been developed have thus all focussed on the *topology* of molecules, without consideration of their *topography*, i.e., the relative orientations of the constituent atoms in 3-D space. However, intensive research over the last few years has resulted in the development of graph-theoretic methods for searching databases of 3-D structures. The resulting systems are used primarily for *pharmacophoric-pattern matching*, i.e., the identification of all molecules in a database that contain some user-defined pattern of atoms in 3-D space: these systems are reviewed by Martin *et al.* [18] and by Willett [24]. The in-house and proprietary systems that have been developed to date have been designed for the processing of small 3-D molecules. In this paper, we discuss an ongoing research project at the University of Sheffield that seeks to develop graph-theoretic methods for the storage and retrieval of the 3-D macromolecules in the Brookhave Protein Data Bank. The organisation of the paper is as follows: Section 2 summarises the main structural characteristics of proteins and discusses the representation of the secondary structure of proteins by labelled graphs; Section 3 describes the design and use of a program, called POSSUM, that uses a subgraph isomorphism algorithm to search for secondary-structure motifs; and Section 4 introduces a second program, called PROTE, that uses a maximal common subgraph isomorphism algorithm to identify structural resemblances between 3-D proteins at the secondary-structure level. The paper concludes with a discussion of areas that need further research.

2 The Structure Of Proteins

2.1 Levels of structural description

The structures of protein molecules can be described at several levels of complexity:

- *Primary structure* is the sequence order of the amino acids along the protein main chain.
- *Secondary structure* involves the use of limited 3-D information about intra-mainchain hydrogen bonds to identify regions of *α-helix*, *β-strand*, or of *turn*, etc. (as discussed below) along the polypeptide chain. However such a secondary structure description says nothing concerning the relative 3-D orientation of these secondary structural elements.
- *Tertiary structure* consists of the 3-D structure of the protein chain: this includes not only information about the relative orientations of secondary structure elements, but also detailed information concerning sidechain orientations, etc.
- Many proteins consist of an assembly of a number of identical or non-identical polypeptide chains: the *quaternary* structure is the description of the way in which these subunits are oriented with respect to one another.

The three main types of secondary structure element (SSE) are as follows:

- The α-helix, which is a tightly-coiled helix (one turn every 3.6 residues) that is stabilized by hydrogen bonds between mainchain carbonyl groups and the mainchain NH group four residues further on in the sequence.
- The β-strand is another regular, repeating structure; here, the main chain is fully extended, rather than tightly coiled as with an α-helix, so that the β-strand appears more like a pleated ribbon structure. Groups of β-strands form β-sheets. A sheet is an example of a *supersecondary structure*, in which groups of β-strands lie next to one another, either parallel or anti-parallel, with hydrogen bonds between mainchain NH and CO groups on adjacent strands.
- There are also turns, which generally link two of the above secondary structures together, *loops* and *random coil* structures, in which there is no discernable regular hydrogen-bonding pattern and the structures are thus much less specific.

The primary source of 3-D coordinate data is that contained in the Protein Data Bank, which functions as the internationally recognized archive of the 3-D structures of biological macromolecules [1,5]. Most of the structures in the Protein Data Bank have been obtained from X-ray single-crystal diffraction studies, but there are also a few structures obtained from neutron-crystallographic, electron-diffraction, NMR and model-building studies. The April 1991 release of the database lists a total of 665 sets of coordinates, with more than 200 further structures that have already been submitted and that should become available within the next two years. It is important to note that many of the depositions in the Protein Data Bank are not unique: for any protein there may be a number of coordinate sets representing different stages of crystallographic refinement, the binding of different substrates to the protein, or different site-directed mutants. Also, in certain cases, different species variants of the same protein may differ in only a few amino acids. It is difficult to quantify exactly, but there are probably less than 300 distinct protein structures in the Protein Data Bank.

2.2 Representation of secondary-structure motifs

Although some of the work that has been carried out in Sheffield has considered searching for patterns of C^{α} atoms [6], our main focus of interest has been the representation and searching of *secondary-structure motifs*, i.e., patterns of SSEs in 3-D space. We have noted previously that the molecules in current database systems for small molecules are represented by labelled graphs, in the form of 2-D or 3-D connection tables, and we have developed an analogous graph-theoretic representation of the secondary structure of a protein. The representation makes use of the fact that the two most common types of SSE, the α-helix and β-strand, are both approximately linear repeating structures, which can hence be described by a vector drawn along their major axes. The set of vectors corresponding to the SSEs in a protein can then be used to describe the structure of that protein in 3-D space, with the SSEs and the inter-SSE angles and distances corresponding to the nodes and to the edges, respectively, of a graph [2,19]. In fact, each edge in such a labelled graph is a three-part data element that contains the angle between a pair of lines, the distance of closest approach and the distance between their mid-points.

The information that is needed to create these labelled graphs is obtained by processing of the coordinate data in the Protein Data Bank, as follows:

1. The α-helix and β-strand SSEs are identified using the Kabsch-Sander algorithm [15].
2. The assignments are checked using the FRODO molecular graphics system [14] and modified if necessary.
3. The residues comprising an SSE are then input to a routine that calculates the major axis of the SSE: this is done by mapping idealized helix or strand structures to the residue coordinates using a least-squares fitting procedure.

4. The axes resulting from Step 3 may be considered as vectors in 3-D space and the final step involves calculating the angles and distances between each pair of axes: these data are then stored in the connection table that is used to represent a protein.

In this way, it is possible to construct a compact database of labelled graphs that can be processed using graph-matching algorithms analogous to those that are used for the processing of small 2-D and 3-D molecules. Specifically, we have used the subgraph isomorphism algorithm due to Ullmann [23] and a maximal common subgraph isomorphism based on the clique-detection algorithm of Bron and Kerbosch [9] as the basis for the programs POSSUM (Protein Online Substructure Searching — Ullmann Method) and PROTE (PROtein Topographic Explorer), respectively. These two particular algorithms have been chosen following detailed comparisons of the applicability of a range of graph-matching procedures to the processing of 3-D chemical structures [7,8]. An alternative approach to the graph-theoretic description of protein structures has been developed recently by Kaden *et al.* [16].

3 POSSUM — use of a subgraph isomorphism algorithm

3.1 Program structure

The POSSUM program was developed to allow searches to be carried out for secondary-structure motifs, i.e., patterns of α-helices and / or β-strands. The Protein Data Bank structures are encoded as described in Section 2 and a similar representation is used for the query motifs. Given such a representation, the identification of a motif in a protein is directly analogous to pharmacophoric-pattern searching in databases of small 3-D molecules, and it is thus possible to use a simple modification of the Ullmann subgraph isomorphism algorithm [23], which is used in several operational 3-D database systems for small molecules [24].

The inclusion of three types of information in the edges of the graphs, i.e., the inter-SSE angles and the two types of inter-SSE distance, provides a range of searching options depending upon the requirements of a particular query, with the precision of the search being controlled by the distance and angular tolerances which the user allows for a match to be present. The distance and angular tolerances can be expressed as absolute quantities, i.e., a certain number of Å or degrees; additionally, the distance tolerances can be expressed as a proportion of the distance concerned, e.g., 10%. The user can also specify whether or not the order that the SSEs occur along the protein chain is important in the search for that motif. Of the two types of distance that can be specified, that of closest approach has been found to be more effective for retrieving secondary structures which interact with each other in some sense, e.g., neighbouring pairs of β-strands in a sheet or α-helices in van der Waals' contact with each other. However, the midpoint distance appears more appropriate for the case where the SSEs are more remote from one another in the structure. Matches to the query pattern are output by the program in a format compatible with the FRODO graphics package; this allows the immediate inspection of the highlighted secondary structures from database proteins that match the query motif, using Digital Equipment Corporation, Evans and Sutherland or Silicon Graphics hardware. The program is written in Fortran 77, and a scan of the entire Protein Data Bank typically requires 4-5 CPU minutes on a MicroVAX-III (though the precise value depends upon the specific motif that is being searched and the error tolerances that are used).

3.2 Searches for motifs

POSSUM has been extensively tested by means of searches of the Protein Data Bank for many different query motifs [2,19]. The results demonstrate that the system provides an extremely efficient means of identifying all occurrences of these motifs, occurrences that, in some cases, had not been recognised previously owing to the complexity of the proteins' structures. In 1977, for example, Richardson reviewed all of the 37 β-strand patterns then known [22]. POSSUM searches were carried out in the Protein Data Bank for 34 of these motifs (the other three not being searched since they are very simple motifs that occur in nearly all strand-containing proteins). Given the extensive studies that have been carried out on the small number of structures in the Protein Data Bank, it is remarkable to find that no less than 15 of the POSSUM searches resulted in the identification of at least one occurrence of the query motif that had not been observed by Richardson [2].

An example of the utility of this approach to protein searching is provided by a recent study that has established a striking resemblance in the tertiary folds of the *Salmonella typhimurium* CheY chemotaxis protein and of the GDP-binding domain of *E. coli* Elongation Factor TU (EF Tu) [3]. The CheY structure contains a five-stranded β-sheet together with five associated α-helices. The ten elements thus give a total of 45 distinct pairs of linear SSEs. In fact, the protein was represented for search by inter-line angles and the distances of closest approach for only one third of this number of pairs of lines, together with the order of the SSEs in the sequence. Although this provides a highly generalised description of the CheY structure, only five of the structures in the April 1989 release of the Protein Data Bank contained the query motif within the specified angular and distance tolerances of 50° and 40%, respectively. Inspection of the retrieved structures revealed a remarkable similarity between the structures of the GTP-binding domain of EF Tu and of CheY, the only major difference being that EF Tu has one extra anti-parallel β-strand at the end of the sheet. The structural relatedness is demonstrated by a least squares superimposition of 64 Cα atoms from the five helices and five strands with an RMS deviation of only 2.4 Å; however, this relatedness is quite unrecognisable from the primary sequences of the two proteins, since a comparison of the amino acids revealed no significant sequence homologies. Previous studies had recognised the resemblances of each protein's fold to that of a generic nucleotide binding domain but the relationship identified by POSSUM is far greater, involving no less than ten SSEs. This structural similarity is most surprising since the two proteins are representatives of two major macromolecular classes that had not previously been thought to be related. CheY is a signal transduction protein with sequence homologies to a wide range of bacterial proteins involved in regulation of chemotaxis, membrane synthesis and sporulation; whilst EF Tu is one of the family of guanosine nucleotide binding proteins which includes the ras oncogene proteins and signal transducing G proteins. POSSUM has thus established a previously unsuspected link between the families of bacterial signal transduction proteins and of signal-transducing G proteins. An independent confirmation of this relationship has recently been reported by Chen *et al.*, who have noted the resemblance between CheY and another GTP-binding protein, the *ras* oncogene protein [10].

3.3 β-sheet topologies in proteins

A widely accepted model of protein folding suggests that the unfolded polypeptide chain rapidly forms marginally stable elements of secondary structure, which then interact to form stabilized units of supersecondary structure or minidomains. There is evidence to suggest that certain structural motifs may be more important than other parts of the molecule in initiating folding. Therefore, analyses of secondary and supersecondary structure motifs may provide important information about the folding process and provide rules or constraints for predicting folding pathways. An ongoing study in Sheffield is using POSSUM to investigate the occurrence characteristics of one important type of supersecondary structure, the β-sheet, in globular proteins.

A sheet comprises a set of β-strands, which are often far apart in the primary sequence. In the following, we use the term β-*motif* to refer both to complete β-sheets and to partial β-sheets that are subsets of larger β-sheets. For identification purposes, all of the β-motifs are described using a binary notation, in which each β-strand in a motif is assigned a 1 or a 0. The first strand in the β-motif is always denoted by 1 and subsequent strands are assigned the values of 1 or 0, depending upon whether they are parallel or antiparallel, respectively, to the first strand. The notation thus represents the relative alignment of the β-strands in the order that they occur in the β-motif (and not their sequence order). In general, for a β-sheet consisting of n strands, there are $2^{(n-1)}$ possible β-motifs, although this total is reduced by considerations of symmetry. For example, a 3-stranded β-sheet can yield the β-motifs 100, 101, 110 and 111; however, only three of these are unique, since 100 can be obtained from 110 (and *vice versa*) by a simple inversion of the relative strand alignment.

A program has been written to generate POSSUM queries for all of the possible β-motifs that can be derived from sheets containing between 3 and 15 strands. The total possible numbers of such motifs are shown in Table 1. A query pattern consists of the angles and distances for neighbouring strands and the distances to second and third neighbour strands. No angular value is set for the next-nearest neighbours because of the common phenomenon of twist occurring among the strands in β-sheets, which makes it very difficult to estimate the characteristic angular values for strands other than nearest neighbours. Consequently, an inter-strand torsion angle of -25° is assigned to neighbouring parallel strands and an angular value of +155°, assigned to neighbouring anti-parallel strands. Generalised closest-approach distances of 4.5 Å, 9.0 Å, and 13.5 Å define first-, second- and third-neighbour strands. Longer-range distances are not defined as this would impose a restraint on the overall curvature of the sheet, and the degree of twist between neighbouring strands along a sheet is seemingly unpredictable and variable from one globular protein to the next. In the searches, we have ignored the sequence order of the strands comprising a sheet, so as to focus on their spatial arrangement in the sheet.

Number of strands (n)	Number of possible motifs $^{(2n-1)}$	Number of possible unique motifs	Number of actual motifs found
3	4	3	3
4	8	6	6
5	16	9	9
6	32	20	15
7	64	36	17
8	128	72	18
9	256	142	11
10	512	288	8
11	1024	576	5
12	2048	1152	1
13	4096	2304	1
14	8192	4608	1
15	16384	9216	1
Totals	**32764**	**18432**	**96**

Table 1: potential and actual occurrences of β-motifs

A typical search matrix, that for the 4-strand motif 1011, is shown in Table 2 over the page. The matrices for all possible motifs containing between 3 and 15 strands were used as queries for POSSUM searches of a subset of the coordinate entries from the February 1990 release of Protein Data Bank. This subset contained a single, high-resolution set of coordinates for 114 distinct proteins. The outputs from these searches thus provide an exhaustive catalogue of all of the occurrences of β-motifs in the Protein Data Bank, and studies are now in progress to analyse the very large amount of data that has been collected.

Strand	1	2	3	4
1	0 0.0	155 4.5	— 9.0	— 13.5
2	155 4.5	0 0.0	155 4.5	— 9.0
3	— 9.0	155 4.5	0 0.0	-25 4.5
4	— 13.5	— 9.0	-25 4.5	0 0.0

Table 2: Search matrix for the β-motif 1011. The first and second figure in each element of the table correspond to the inter-line angle and to the closest-approach distance, respectively

It can be seen from Table 1 that a vast number, 18432, of β-motifs are feasible, for $3 \leq n \leq 15$, but that only 96 (just over 0.5%) were actually found to be present in the Protein Data Bank structures. In general, the smaller motifs are retrieved from a greater number of proteins than the larger motifs (as would be expected as many of the smaller motifs are subsets of larger β-motifs). Small motifs that are less prevalent in the Protein Data Bank structures occur less frequently as subsets of larger motifs, or in some cases are completely non-existent. For example, the motif 100011 was identified only in arabinose-binding protein and was not retrieved as a subset of any larger β-motifs. It is not clear why only a small number of the possible β-motifs occur and why the vast majority do not occur (although it should be remembered that there are still only a small number of structures in the Protein Data bank and many of the larger, unobserved motifs may indeed exist). All of the possible β-motifs occur at least once up to and including the 5-stranded motifs. However, as shown in Table 1, one-quarter of all of the possible 6-stranded motifs were not found, and the fraction of undetected motifs rises extremely rapidly thereafter. Detailed study suggests that there are two primary classes of non-existent motif. The first disfavoured group seems to comprise adjacent units of three or more parallel strands with each of the units in antiparallel alignment, e.g., the motifs 111000 and 1000111. The second disfavoured group appears to be composed of a unit of three or more strands that are parallel to each other, immediately adjacent to a unit of three or more strands in which the strands are antiparallel to each other, e.g., 1000101. The availability of an exhaustive listing of all of the β-motifs in a protein provides a basis for the calculation of quantitative measures of resemblance between pairs of proteins (as evidenced by their constituent motifs). The resulting measures are now being used for the automatic identification of groups of structurally-related proteins and thence for comparison with manually-derived groupings of proteins.

The analysis is now being extended to consider the connectivities of adjacent strands in each of the motifs that have been identified. This work will be based on the study by Richardson [22], who has provided a compact and precise way of describing the mainchain connections of the sequence of strands that comprises a β-sheet. The necessary connectivity information can be extracted automatically from the Protein Data Bank files, and allows the generation of a more detailed characterisation of the structure of a β-motif than that used so far, since motifs with the same structure when using the {0, 1} notation described above may have differing connection patterns. In the longer term, we hope to compare β-sheet topologies at the primary structure level, using sequence comparisons to complement the graph-based relationships identified by the POSSUM searches.

4 PROTE — use of a maximal common subgraph isomorphism algorithm

POSSUM is based on the use of a subgraph isomorphism algorithm with the linear α-helix and β-strand characterisations. PROTE is a more recent program that is based on the use of a maximal common subgraph isomorphism algorithm to identify patterns of α-helices and β-strands in 3-D space that are common to a pair of proteins. The algorithm can be used to identify just the largest common pattern (and hence the term 'maximal common subgraph isomorphism') but is used here to identify all common patterns that are larger than some user-defined threshold size [12]. This restriction is imposed to ensure that the user is not overwhelmed with a very large output consisting primarily of small common substructures that are of little structural significance: even so, a PROTE search can often result in the identification of large numbers of matching substructures that need to be inspected on a graphics terminal. As noted previously, the particular algorithm used is the clique-detection procedure originally devised by Bron and Kerbosch [9].

PROTE has been tested using sets of structurally related proteins and has been shown to be capable of identifying correctly the areas of structural commonality. For example, the myoglobin structure 1MBD, which contains nine α-helices, was used as the target protein for a search of all of the 371 structures in the April 1989 release of the Protein Data Bank. This release contained 38 examples of other oxygen-carrying globin structures: 16 hemoglobins, 8 myoglobins and 14 plant leghemoglobins. Although there is a wide variation in the sequence homology between such proteins, they all possess very similar 3-D structures. Myoglobins are single-subunit molecules whereas hemoglobins consist of several globin-type subunits and the plant leghemoglobins have small differences in steric positioning of residues around the ligand-binding site but are essentially very similar in structure to other globins.

Sets of runs of PROTE with 1MBD as the pattern protein were carried out with the minimum output clique size set at a value of 6, i.e., only common substructures containing at least six matched SSE's were output. In the searches reported here, it was required that the cliques in the database structures that matched those in the query protein should have the same sequence ordering of SSEs (though it is also possible to remove this constraint when searching).

The user can specify angular and distance tolerances prior to commencing a search, in just the same way as with POSSUM, to control the numbers and the quality of the matching cliques that are found. In this investigation sets of runs were done where the angular tolerance was varied between 10° and 50° in 10° intervals and the distance tolerance was varied between 10% and 50% in 10% intervals and also between 1 Å and 4 Å in 1 Å intervals. The combination of these possible tolerances with the two types of inter-line distance (closest-approach or midpoint) gave a total of 90 different parameter settings that needed to be tested.

An examination of the cliques of size ≥ 7 located during the set of 90 runs showed that they comprise hits in all of the 16 hemoglobins and 8 myoglobins in the Protein Data Bank, with no hits being found in any other types of proteins. A similar examination of the cliques of size-6 located during this set of runs showed that there are also additional hits in all of the 14 leghemoglobins but that there are, again, no hits in any other types of proteins. All the cliques of size ≥ 7 are 'correct', i.e., the α-helices in the pattern protein 1MBD match with spatially-equivalent helices in the database globin proteins. In the closest-approach distance runs, cliques are found in all of the myoglobin proteins and appear together at tolerances of 10°/30%,3 Å (i.e., using an angular tolerance of 10° with distance tolerances of 30% or 3 Å). Hemoglobin cliques start to appear at 20°/20% and 10°/4 Å. At 30°/50% cliques have been found in 23 out of the 24 hemoglobins and myoglobins (no cliques are found in the hemoglobin 1HDS in any of the closest-approach/% runs); at 30°/4 Å, cliques have been found in all 24 of the hemoglobins and myoglobins. Similar results are found in the midpoint distance runs. Cliques are found in all of the myoglobins and appear together at tolerances of 10°/30%,3 Å. Hemoglobin cliques start to appear at 10°/30% and 20°/3 Å; at tolerances of 20°/30% and 40°/4 Å cliques are found in all 24 of the hemoglobin and myoglobin proteins. However, some 'incorrect' matches of helices in the pattern and structure proteins are found when high tolerance settings are used, e.g., 50°/50% for 1HDS.

The size-6 matches in this set of runs include cliques found in all the 14 plant leghemoglobins, as well as additional cliques in some of the hemoglobin and myoglobin proteins. Some 'incorrect' matches of helices in the query and database proteins were found in high-tolerance midpoint distance runs (50°/40-50%) for two hemoglobins, six myoglobins and all 14 of the leghemoglobins.

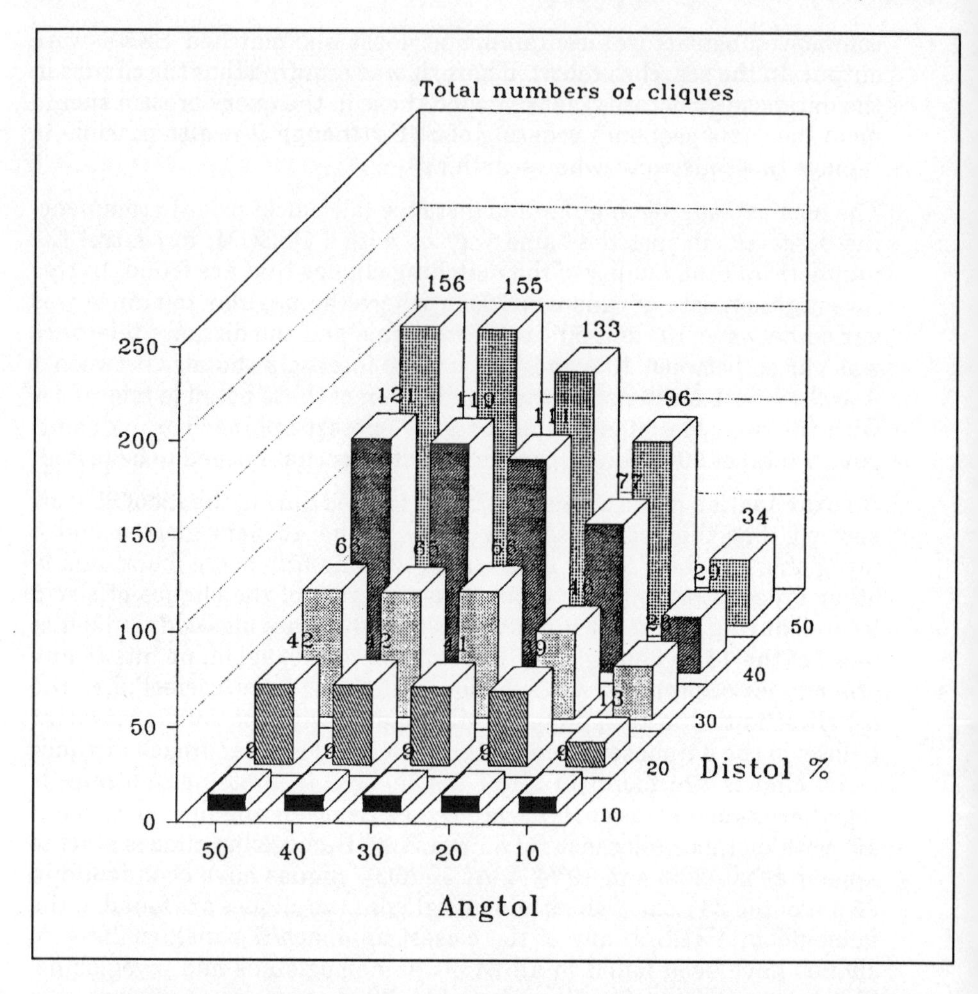

Figure 1: Total numbers of cliques containing at least six SSEs found in PROTE searches that used 1MBD as the target protein. The results correspond to a closest-approach distance search in which the tolerances are expressed as a percentage of the closest-approach distance in the target. The numbers attached to each bar denote the number of cliques that were identified using each combination of search parameters.

The results of the runs are conveniently displayed by means of a 3-D bar chart, such as that shown in Figure 1, where the total numbers of cliques found in the runs for each set of tolerances are indicated by the height of the relevant bar in the bar chart. This particular figure summarises the results that were obtained in searches that used the closest-approach distance, with the tolerances expressed as a percentage of the distances specified in the target protein; similar figures may be obtained for the other types of search. An inspection of the bar charts reveals that the numbers of cliques found greatly exceeds 38, which is the total number of globin proteins in the Protein Data Bank, especially at the higher tolerances. This is because more than one clique may be located within a database globin structure. These cliques may comprise 'correct and 'incorrect' cliques and may be of a size smaller than the total size of the pattern protein, hence enabling several unique but overlapping cliques to be found. The phenomenon of clique overlap may be illustrated by noting that several cliques of size-7 found with the pattern protein and a database structure may, for example, contain six matched SSEs which are the same, but differ in the identity of the seventh matched SSE.

In conclusion, quite low tolerances produced the expected cliques in known globin proteins for closest-approach and midpoint distance runs, using a minimum match size of 6 SSEs and using both % and Å distance tolerances. In the % tolerance runs, at least one clique in every one of the 38 globin proteins in the database was found where:

- The angular tolerance was ≥ 20, AND
- The distance tolerance was ≥ 50% (closest-approach) or ≥ 40% (midpoint).

In the Å tolerance runs, cliques were not found in all of the 38 globin proteins. The closest-approach distance runs produced cliques in 35 of the globins, and midpoint distance runs produced cliques in the 24 hemoglobins and myoglobins. These hits were obtained if:

- The angular tolerance was ≥ 20, AND
- The distance tolerance was ≥ 4 Å.

The parameter values described here are those necessary for effective retrieval with 1MBD as the target structure in a search for globins; other values may be appropriate for other types of search, and we are now completing an extended series of searches (using a wide range of types of query pattern) to determine appropriate default search parameters.

The globin results, and many others, indicate that the clique detection program is effective in finding matches with other proteins of similar 3-D structure, where all of the structures belong to the same family of proteins. While it is clearly of interest to be able to identify such structural resemblances automatically, rather than via the manual inspection of a graphics terminal, the program has been developed for the automatic identification of areas of structural commonality that had not previously been detected. Our current work hence involves taking some particular protein structure and then using this for a scan of all of the proteins in the Protein Data Bank to identify the substructures that the query structure has in common with each of the other members of the database; in essence, this provides a mechanism for the measurement of the degree of resemblance between the query and all of the other proteins. In addition, if some particular common substructure is found to occur with several different structures from the database, this might suggest the presence of a novel, previously-unidentified, secondary-structure motif.

5 Conclusions

When the Protein Data Bank was first set up in the late Seventies, it was intended to act as an archive for the deposition of sets of macromolecular coordinates. The burgeoning demands of research in molecular biology and biotechnology has led to increasing use being made of the information that is stored in databases of 3-D chemical substances; in particular, much attention has been devoted to the development of sophisticated molecular graphics systems. In this paper, we have demonstrated that this data can also be used for a range of searching operations, using relatively small modifications to well-established techniques that were first developed for the storage and retrieval of small chemical structures. To date, we have considered only the linear α-helix and β-strand SSEs but will shortly extend our methods of representation to encompass other types of structural features, such as turns, loops, metal and substrate binding sites and disulphide bridges; in the longer term, it may prove possible to integrate searching at the secondary structure level with searching at the residue level, using either sequence data (as suggested at the end of Section 3) or the C^{α} coordinates (using techniques discussed by Brint *et al.* [6]. There are, of course, other approaches to the searching of 3-D protein structures (see, e.g., [11,13,20,21]) and these might usefully be combined with the graph-theoretic approaches that have been developed in Sheffield and those that have been developed by Kaden *et al.* [16]. In the longer term, one may expect a trend towards integrated systems that provide a common interface and searching capabilities for both small 3-D molecules and 3-D macromolecules.

Acknowledgements

Our work on 3-D protein searching is currently supported by Pfizer Central Research, the Science and Engineering Research Council and Tripos Associates. Peter Artymiuk is a Royal Society University Research Fellow and David Rice is a Lister Institute Research Fellow.

References

[1] Abola, E. E., Bernstein, F. C., Bryant, S. H., Koetzle, T. F. & Weng, J. (1987). (Protein Data Bank). In: Allen, F. H., Bergeroff, G. & Sievers, R. (eds.), 'Crystallographic Databases: Information Content, Software Systems, Scientific Applications'. Cambridge: Data Commission of the International Union of Crystallography.

[2] Artymiuk, P. J., Mitchell, E. M., Rice, D. W. & Willett, P. (1989). Searching techniques for databases of protein secondary structures. *Journal of Information Science*, **15**, 287-298.

[3] Artymiuk, P. J., Rice, D. W., Mitchell, E. M. & Willett, P. (1990). Structural resemblance between the families of bacterial signal

transduction proteins and of G proteins revealed by graph theoretical techniques. *Protein Engineering*, **4**, 39-43.

[4] Barnard, J. M. (1989). Recent developments in chemical structure handling. *Perspectives in Information Management*, **1**, 133-168.

[5] Bernstein, F. C., Koetzle, T. F., Williams, G. J. B., Meyer, E. F. Jnr., Brice, M. D., Rodgers, J. R., Kennard, O., Shimanouchi, M. & Tasumi, M. (1977). The Protein Data Bank: a computer-based archival file for macromolecular structures. *Journal of Molecular Biology*, **112**, 535-542.

[6] Brint, A. T., Davies, H. M., Mitchell, E. M. & Willett, P. (1989). Rapid geometric searching in protein structures. *Journal of Molecular Graphics*, **7**, 48-53.

[7] Brint, A. T. & Willett, P. (1987). Algorithms for the identification of three-dimensional maximal common substructures. *Journal of Chemical Information and Computer Sciences*, **27**, 152-158.

[8] Brint, A. T. & Willett, P. (1987). Pharmacophoric pattern matching in files of three-dimensional chemical structures: comparison of geometric searching algorithms. *Journal of Molecular Graphics*, **5**, 49-56.

[9] Bron, C. & Kerbosch, J. (1973). Algorithm 457. Finding all cliques of an undirected graph. *Communications of the ACM*, **16**, 575-577.

[10] Chen, J. M., Lee, G., Murphy, R. B., Brandt-Rauf, P. W. & Pincus, M. R. (1990). Comparisons between the three-dimensional structures of the chemotactic protein CheY and the normal Gly 12-p21 protein. *International Journal of Peptide and Protein Research*, **36**, 1-6.

[11] Gray, P. M. D., Paton, N. W., Kemp, G. J. L. & Fothergill, J. E. (1990). An object-oriented database for protein structure analysis. *Protein Engineering*, **3**, 235-243.

[12] Grindley, H. M. (1991). PhD thesis, University of Sheffield, in preparation.

[13] Islam, S. A. & Sternberg, M. J. E. (1989). A relational database of protein structures designed for flexible enquiries about conformation. *Protein Engineering*, **2**, 431-442.

[14] Jones, T. A. (1985). Interactive computer graphics: FRODO. *Methods in Enzymology*, **115**, 157-171.

[15] Kabsch, W. & Sander, C. (1983). Dictionary of protein secondary structure: pattern recognition of hydrogen-bonded and geometrical features. *Biopolymers*, **22**, 2577-2637.

[16] Kaden, F., Koch, I. & Selbig, J. (1990). Knowledge-based prediction of protein structures. *Journal of Theoretical Biology*, **147**, 85-100.

[17] Lipscombe, K. J., Lynch, M. F. & Willett, P. (1989). Chemical structure processing. *Annual Review of Information Science and Technology*, **24**, 189-238.

[18] Martin, Y. C., Bures, M. G. & Willett, P. (1990). 'Searching databases of three-dimensional structures'. In: Lipkowitz, K. B. & Boyd, D. D. (eds.), Reviews in Computational Chemistry. New York: VCH.

[19] Mitchell, E. M., Artymiuk, P. J., Rice, D. W. Willett, P. (1990). Use of techniques derived from graph theory to compare secondary structure motifs in proteins. *Journal of Molecular Biology*, **212**, 151-166.

[20] Rawlings, C. J., Taylor, W. R., Nyakairu, J., Fox, J. & Sternberg, M. J. E. (1985). Reasoning about protein topology using the logic programming language PROLOG. *Journal of Molecular Graphics*, **3**, 151-157.

[21] Richards, F. M. & Kundrot, C. E. (1988). Identification of structural motifs from protein coordinate data: secondary structure and first-level supersecondary structure. *Proteins: Structure, Function, and Genetics*, **3**, 71-84.

[22] Richardson, J. S. (1977). β-sheet topology and the relatedness of proteins. *Nature*, **268**, 495-500.

[23] Ullmann, J. R. (1976). An algorithm for subgraph isomorphism. *Journal of the ACM*, **16**, 31-42.

[24] Willett, P. (1991). 'Three-Dimensional Chemical Structure Handling'. Taunton: Research Studies Press.

Catalyst: a computer aided drug design system specifically designed for medicinal chemists

P.W. Sprague

BioCad Corporation, 1091 North Shoreline Blvd., Mountain View, California 94043, USA

Introduction

Modern drug discovery efforts in pharmaceutical companies place unusual demands on automated information handling systems. Drug discovery programs attempt to identify molecules in the 300-700 Daltons molecular weight range that have specific activities of interest, and acceptable pharmaceutical properties for development into marketable drugs. With today's emphasis on 'rational' drug design, projects proceed through several distinct phases including conceptualization, lead identification and optimization, database searches and document preparation. There will also be troubleshooting exercises required by the occurrence of unexpected problems along the development path. In this paper, we will consider each phase along the drug discovery path and discuss a new set of software tools that have been specifically designed to facilitate drug discovery work.

Problem Conceptualization

Rational drug discovery programs often begin with a collaboration between medicinal chemists and scientists from one of the life sciences such as pharmacology. Typically, new facts appear in the literature which suggest, to these individuals, a novel approach to disease control. For example, publication of work by Ferreira on bradykinin potentiating peptides isolated from Bothrops jarara prompted Ondetti and Cushman at Squibb to investigate these substances as a new approach to treating high blood pressure.[1,2] Isolation of the active peptides from the snake venom and proof of their structure by synthesis gave the first detailed molecular information from which efforts towards small molecule inhibitors could be rationally launched. [3]

At this point, energetically reasonable three dimensional structures of these inhibitors would have been very valuable for planning small molecule synthesis. In conceptualizing interactions between molecules, it is useful to assume that the interacting molecular partners adopt conformations which facilitate the developing interaction; a sort of 'induced fit' phenomenon. Such conformations should be energetically reasonable so the overall cost in energy does not render the molecular interaction unfeasible. While it is possible to intuit these conformations from physical models when structures are very rigid, most molecules possess so many acceptable conformations that the 'Dreiding model' approach becomes too complex to be practical. This is where a computer system capable of dealing with thousands, or perhaps millions, of conformations for each molecule would be welcome.

Beyond visualization, there is the problem of developing a good hypothesis based on the limited structure and activity data set available at an early stage in a new drug development project. Usually, this is done by relying entirely on intuition, including whatever bias the author may have. A method that could impartially consider the three dimensional arrangement of important molecular binding features, would be an important adjunct to the medicinal chemist's intuition.

At BioCAD, we have developed a software system named Catalyst that possesses this and other capabilities. Molecules are drawn with a two dimensional drawing editor that works in concert with a three dimensional rendering program. Images are simultaneously presented as two and three dimensional objects. These structural renderings are interactive and interrelated; changes in one representation are instantly updated in the other. Furthermore, when the 3D images are manipulated on the screen, as with a ring flip operation for example, real time calculations are carried out in the background to quantitate the changes in strain energy imposed on the molecule. As you bend a ring, the entire structure adjusts conformationally in real time. Thus, Catalyst's 3D models are dynamic because they give the user a feel for energetics as well as the 3D shape of a molecule.

Interactions between 3D molecules occur primarily between functional elements (features) such as charges, hydrogen bond donors, acceptors, etc. Catalyst identifies all potential binding features in the set of molecules under analysis. Relative feature sets are ranked according to the relative biological activity of the molecules, identifying features which favor, and disfavor, activity. Catalyst also considers the 3D positioning of each feature set by examining the conformational space available for each molecule. The program then models the set of features whose arrangement in 3D space best predicts activity. The result of this analysis is displayed as an object, rendered in three dimensions, that depicts the most important structural features affecting activity, and to which we have given the name hypothesis.

Hypotheses can be automatically regenerated or interactively edited to include the effects of new biological data and to reflect new constraints if desired. They can also serve as queries for searching a database of 3D structures for compounds that fit the constraints. Catalyst includes very fast 3D searching that makes full searches of very large databases (millions of structures) practical. Because the search is biased only by the constraints of the hypothesis, and there is no limit on the structural diversity permitted in the database, this process can lead to the identification of novel, sometimes, quite unexpected molecules. These compounds, or perhaps pieces of them, can stimulate the imagination of medicinal chemists and become starting points for synthesis work.

Lead Identification

The 3D hypothesis paradigm is a powerful tool for focusing synthetic efforts to identify lead compounds. Lead compounds are generally defined as those having specific activity for the target enzyme or receptor, and *in vitro* potency, expressed in IC_{50} terms, in the single digit micromolar range. Medicinal chemists are often presented with compounds whose activity profile encompasses the 10 to 1000 μM range. The goal then becomes identifying a derivative with a 10 fold increase in potency. If the set of molecules used to derive a Catalyst hypothesis represents enough structural diversity to bracket the optimum active structure, Catalyst can identify that optimum. With the Catalyst hypothesis generator and an appropriate structure activity data set, computer assisted identification of the best possible synthetic target molecules becomes a practical exercise.

Access to an hypothesis generator will modify the synthetic strategy employed by chemists. A hypothesis will be most useful when a rich variety of structures is used. The chemist can deliberately assure that this situation exists by careful selection of early synthetic targets. Thus, examination of broad areas of structural space will be efficient and lead quickly to the most promising structural classes.

Lead Optimization

Once chemists have learned how to make a compound with specific activity and potency in the 1 μM range, the goal shifts to lead optimization. Typically, a potency of IC_{50} = 1-10 nM is desirable to assure enough activity for a marketable oral agent, and finding such a compound becomes the project goal. The approach often taken is to synthesize every chemically possible derivative structurally related to the lead. While sometimes successful, this approach is very manpower intensive and often fails to identify compounds 100-1000 fold more active than the starting lead.

The Catalyst hypothesis generator will have a major impact on how the optimization process is carried out. Prospective synthetic targets can be quantitatively evaluated for fit to the hypothesis before they are actually synthesized. Only the most promising compounds need actually be prepared and tested.

The program will also help chemists discover unexpected structural motifs in the event that the project goal activity cannot be achieved by following the current lead. This is possible because a Catalyst hypothesis can be used as a query to search 3D structure databases. Given a large database with a rich variety of structural types, the process can produce structures unbiased by human subjectivity. This process is similar in kind to a natural products search, where large numbers of diverse structures are searched for a specific activity. Such methods have historically been very successful in producing novel lead compounds. Catalyst makes this kind of approach possible in the virtual sense for the first time.

Databases

Any modern drug discovery program quickly generates large amounts of data, both structural and alphanumeric. If this data is to be useful for planning, a robust and convenient mechanism must be provided for storage and retrieval. Database systems commonly in use today rely on database engines mounted on remote mainframe computers providing users access through terminals. Many of these programs lack a friendly interface or run in very busy environments so that operation and response times are disappointingly slow. Sending queries to these databases usually results in one of two unsatisfactory results. First, the response may be quite precise, delivering exactly the answer desired, but take so long that the user becomes frustrated. Alternatively, the response may be quite rapid, but the number of answers so large that the user is quickly overwhelmed by the data.

Ideally, users should have precise answers quickly. Data storage and retrieval must be tightly coupled to the drug discovery process. Chemists must be able to query even very large structure collections interactively. The database engine in Catalyst was constructed with all of these concerns as design goals. Thus, response times are very fast and there has been no sacrifice in answer quality. 3D searches of compound databases containing hundreds of thousands of records are accomplished in seconds on a desktop workstation.

The Catalyst database tools include provisions for importing structures in batch mode from existing corporate archival databases. There are also facilities for including alphanumeric data (biological data, for example) integrated with the structures.

Report Generation

Modern medicinal chemistry labs now include personal computers for each chemist along with the more traditional equipment. These computers are used for office automation tasks, such as electronic mail, word processing and preparation of presentation materials. Medicinal chemists in the pharmaceutical industry are often called upon to prepare reports of their results, and these tools greatly facilitate those tasks. Catalyst provides a full set of annotation tools for preparing structure tables. In addition, there is 'cut and paste' compatibility with third party word processors, drawing packages and spreadsheet programs. To ensure user acceptance, the Catalyst user interface has been deliberately designed to be intuitive and easy to use.

Summary

Catalyst is the first integrated drug discovery software package specifically designed for medicinal chemists. It includes new tools, designed from first principles, for drawing and rendering molecules in two and three dimensions. Catalyst contains novel hypothesis generation tools for extracting the important information from data sets of structures and biological activity. The resulting hypothesis object can be used to search for active structures in 3D structure databases. New database storage and retrieval methods have been designed specifically to address problems in dealing with very large chemical compound and biological activity databases. Finally, Catalyst has been integrated with a user friendly, intuitive interface and a variety of office automation software tools.

References

[1.] S. H. Ferreira, D. C. Bartelt and L. J. Greene, *Biochemistry*, **12**, 2070, 1973.

[2.] M. A. Ondetti, N. J. Williams, E. F. Sabo, J. Pluscec, E. R. Weaver and O. Kocy, *Biochemistry*, **10**, 4033, 1971

[3.] M. A. Ondetti, B. Rubin and D. W. Cushman, *Science*, **196**, 441, 1977.

Infometrics for mapping and measuring science and technology

W.A. Turner

CERESI/CNRS, 26 rue Boyer, 75020 Paris, France

Abstract

The structure of the information industry considerably changed during the 1980s. Of particular interest to us in this paper is the desk-top packaging of information for use in local work contexts. Networks, micro-computers and software developments suggest the need for a 'distributed' approach to information management. Approaches of this sort are currently being experimented in industry. We intend in this paper to report on one such experience.

1) Managing information flows

Chemco is a large multinational chemical company. Its Information Service has systematically collected the 'in-house literature' of the company (reports, publications, patents, notes, agreements . . .) and downloaded from external host services information concerning its various sectors of scientific and technological activity. The files can be accessed from any micro-computer in the company's network and have been regularly updated since the beginning of the 1980s.

The Company's information management system is a well-known high performance standard product. Documents in the system are indexed by means of an open, non-controlled lexicon. This lexicon is built automatically by application of a simple character string recognition algorithm. All the 'words' in the lexicon are uniterms. A uniterm is defined as an independant character string identified by two markers that precede and follow it. These markers can be, for example, a blank space or a predefined punctuation mark such as the comma, semicolon, period,etc. A uniterm will be included in the lexicon only if it does not appear on a proscribed stop word list of non-relevant indexing terms. A file which lists and localizes all the appearances of a given uniterm in the document set is generated. This file is used for document retrieval purposes. A user of the system can use proximity measures, truncature commands and boolean operators to combine uniterms in order to formulate his or her search strategy.

Another tool recently acquired by the Information Service serves for reformatting and detecting duplicates in document files downloaded from host services. One of the most frequent and time consuming tasks when building an exhaustive 'in-house' file for monitoring science and technology is that of integrating heterogeneous information into a single common format. A company needs to diversify its information sources. No two databases offer exactly the same coverage of science and technology. Furthermore, in-house files generally have to be 'custom tailored' before dispatching information to the various workstations in the network. This means packaging information upon request. The sought-after flexibility in data management can only be achieved through the adoption of a common format.

The goal of custom-tailoring information packages has lead the Information Service to define three types of services. The first has traditionally been a SDI service (Selected Dissemination of Information). The Information Service has a stock of standard queries which it uses to update the 'personalized' information profiles sent out to various Departments and Services of Chemco each month. These queries represent the users' view of their information needs. Initially the profile was sent out on paper, however, the Information Service is making increasing use of electronic mail to address the profile directly to the appropriate mailbox of the Chemco network. The Departments are happy with this solution because the data can be directly 'plugged' into software packages for local secondary analysis.

This move towards electronic delivery of 'personalized' information could have an impact on the division of labour in the Chemco network. The traditional SDI service was offered on the assumption that end users would ultimately be responsi ble for quality control checks and eliminate the unwanted documents on their profiles. This assumption no longer seems appropriate in the electronic mail context. End-users want to use data to test their ideas; to explore different correlations between variables; to build models. Spread-sheets are already being used to achieve these different objectives. More powerful tools are becoming available as a result of research aimed at developing decision-support systems [1]. However, it has become clear that end-users are not specifically interested in acquiring the necessary technical skills needed to build the databases needed for their modeling exercises. Their confidence in the data is based upon their capacity to consult and understand the 'benchmarks' used to measure the quality of the information circulating in the network. The Information Service's new role is rapidly becoming that of building and applying these benchmarks for information flow management.

A second service offered by Chemco's Information Service lies in the field of top-down policy planning exercises. The goal is generally one of helping management to evaluate the technological competence of the firm given discussions about the risks and opportunities of a particular course of action. In recent years, this notion of technological competence has been added to the list of factors that must be analyzed in order to position a firm with respect to its competitive environment. Traditionally, this list was restricted to market considerations, the behavior of competitors and costs. But now in the midst of rapid scientific and technological change, a firm is increasingly obliged to study ways of diversifying existing technologies and of developing new ones [2]. The Information Service of Chemco is periodically called upon to gather information that might help in making decisions of this kind. Generally, scientific and patent databases are used to satisfy this type of request. Classification codes or a key-word formulation of user-needs are used to identify appropriate categories for statistical analysis. Each category represents a subject area that reflects a breakdown of Chemco's technological competence in the field being studied for possible future action. The documents falling in each category are sorted according to a variety of criteria. Publication and patent counts are often used to compare Chemco's activity in each area with that of its competitors and to study changes in the relative importance of this activity over time.

The final service for which Chemco sought our outside assistance was in connection with the development of bottom-up planning exercises. Bottom-up as opposed to top-down planning exercises are not based upon an *a priori* definition of management needs. *A priori* definitions tend to focus attention on existing know-how: the world of changing science and technology is viewed through the distorting glasses of today's knowledge base. Bottom-up planning exercises attempt to overcome this problem. In the next section of this paper we will discuss this particular approach to the general problem of monitoring science and technology.

2) Building a scientific and technological alert system (a STAS)

How can an Information Service become the driving force behind bottom-up planning exercises? The goal is to alert management as to possible risks and opportunities that need its attention given changes in the scientific and technological environment of the firm. In concrete terms this means building a scientific and technological alert system (a STAS). Three questions are of particular importance in this context.

The first is the general question of determining what is significant and what is not in the flow of information entering a network. As we have just said, bottom-up planning exercises are characterized by the fact that department heads, group-leaders and service directors are not asked to specify their particular centers of interest. The rule of the game is 'no *a priori* definition of user needs'. But without such a definition, what benchmarks can be used to determine relevant information for the STAS?

The second problem is that of visualizing the impact of new information on the existing knowledge base. The STAS concept implies that incoming information cannot simply be accumulated and stocked for eventual use by members of the Chemco network. Its impact on existing knowledge structures must be measured.

The third problem derives from the specific nature of planning exercises. Planning means preparing tomorrow on the basis of today's information about trends and resources. An Information Service can produce measures aimed at evaluating a firm's technological competence given its competitive environment. However, its up to management to include these measures in discussions about appropriate resource allocation strategies. The STAS concept implies a clear division of labour. An Information Service can propose the need for action, but in the end it is up to management to decide. An Information Service must consequently win the support of a spokesman in management circles if an early warning system is to work efficiently. A spokesman is somebody who has the ability of initiating collective action after having been alerted to the need to rethink resource allocation strategies in a given sector of activity. Generally, a spokesman will be a department director, a head of service, a group or project leader. This need to 'find a spokesman' raises two problems for the Information Service.

The first is that of deciding where to send its information in the network for evaluation and action. Who is the person that should be alerted to changes in the competitive environment of the firm? Evaluating the need for change in a knowledge base structure essentially means addressing questions which have not yet received attention by the organization. Which group, service or department should be consulted for advice and action when responsibility for acting in a new area has not yet been officially established? Dispatching information to the wrong mailbox implies several costs. The first is that the message might not be understood. The second is that the receiver (department, group, or service) will not act because the perceived risk or opportunity is considered to lie outside the scope of its legitimate field of action. The third is to discredit the STAS concept because if early warning messages are sent out and nobody acts upon them, the role of the Information Service as a driving force in bottom-up planning exercises will be seriously compromised.

However, even when the appropriate person is identified in the network, it is still not certain that he or she will call into question existing practices. A potential spokesman for a new resource allocation strategy needs arguments to defend the idea of strengthening, diversifying or radically changing the technological competence of a firm. This implies a clear understanding of the benchmarks and rules used by the Information Service to pinpoint the need to act in a given area.

These different observations lead to the following conclusion: building a STAS implies fine-tuning a detector device to monitor the information flow. The detector device should have two functions. The first is to collect information and measure its importance; the second is to develop models that will help in drawing conclusions about the risks and opportunities of the competitive environment. We use the general notion of *infometrics* to describe this work of building a detector device to monitor the information flow.

In the rest of this paper we will describe our general approach to infometrics in the Chemco context. Our discussion is organized in three parts corresponding to the different problems presented above.

Our work aimed at detecting the significant elements in an information flow builds upon an approach to natural language analysis that has been developed and tested in a variety of contexts. We have consequently had no great difficulty in developing techniques for measuring the information flow in the Chemco context.

The problem of dispatching information to various points in a distributed network is a new one for us. Much of the work we've done thusfar has been purely exploratory. We will consequently limit our discussion of the dispatching algorithm to a fairly theoretical presentation of the type of solution we are seeking to adopt for this second problem on our work programme

Finally, the general problem of measuring the impact of information flows on existing knowledge structures is a question which we have studied in connection with research into techniques for mapping the dynamics of science and technology. However, this is a particularly difficult problem. We will nevertheless present an example of some of the results thus far obtained in the Chemco context.

3) Measuring the information flow

The Information Service of Chemco has been storing an 'in-house' collection of documents in a variety of electronic files for over the last 10 years. Each department in Chemco is characterized by a folder which traces the history of its activity over the period considered. Among the various files in a folder, there are patents, publications, reports, notes, letters, projects etc. The information content of these different files is not the same. For example, the files containing notes, letters, projects and reports resume to a certain extent the informal discussions that have taken place over time in a given department, whereas the files of publications and patents give a formal description of results obtained. The informal information sources can be used to consider resource allocation strategies which might be appropriate for action at a given moment in time. The patents and articles can be used to draw conclusions about resource allocation strategies which have actually been adopted. This distinction between formal and informal information sources was used to code the information extracted from a folder as we will see later.

The different 'in-house' folders managed by the Information Service were used for building benchmarks to monitor the information flowing through the Chemco network. A benchmark is defined as the list of key-words extracted semi-automatically from a department's document folder at a given moment in time ('t'). The meaning of 'key-word' and 'semi-automatic extraction' needs to be explained in order to fully understand this definition.

A key-word is not a uniterm. The latter has already been defined above as a simple character string which generally appears between two blank spaces in a document. Relevancy is determined by reference to a proscribed stop-word list. All character strings are considered as being potentially relevant for document retrieval if they do not appear on the stop-word list. The establishment of a key-word list requires the use of more stringent quality control measures. Much research has been carried out on this question since the mid 1960's, when G. Salton started his SMART project [3]. Both statistical and natural language processing techniques have been used [4]. Our approach has been described elsewhere under the heading of LEXINET [5].

3.1) LEXINET: a lexicon management tool

LEXINET seeks to combine both natural language and statistical methods in a four step process:

a.) Algorithms have been developed in order to normalize uniterm character strings. They use a predefined list of suffixes and a set of rules designed to identify word stems. These algorithms have thus far been tested for French and English and work has started on Spanish [6]. The goal of grouping together character strings which have the same word stem is two-fold. The first is to reduce the size of the uniterm lexicon which is automatically generated by the standard information management systems which are now available on the market and which we described above when talking of the one used by Chemco. As we will see in a moment, the size of a uniterm lexicon determines the computing time needed to identify compound words. The bigger the lexicon the more time consuming and costly is the computer processing carried out. The second goal is one of quality improvement: the word stem normalization algorithm eliminates sources of statistical error when testing for the existence of compound words.

b.) Compound words are defined as word stems which systematically appear together in a document file in an association sequence ordered from left to right. The statistics used to calculate this sequence are provided by the information management system. A standard IMS automatically generates a file for each uniterm listing its locations in the analyzed corpus. In other words, this file can be used to calculate the number of times a given uniterm appears in the file (C_i) and the number of times it appears in an ordered association sequence with another uniterm (C_{ij}). We have tested several coefficients and have chosen the following as being the most appropriate for identifying compound words [7]:

$$A_{ij} = \frac{C_{ij}^2}{C_i \times C_j} \quad [1]$$

This association coefficient is used to test for the number of times the word 'i' appears in a sentence *before* the word 'j' given their overall frequency in the total file. The value of A_{ij} will be highest when 'i' and 'j' often appear together in ordered association sequences in the corpus. These values are normalized to eliminate the statistical bias introduced by the individual occurrences of each uniterm in the global file.

c.) The result of steps 1 and 2 is a word list which now contains both uniterms and compound words. A final automatic quality control test is carried out before this list is presented for manual validation. This test measures the usefulness of each word on the new list for document indexing. Usefulness is defined in terms of the word's variance:

$$Var(C_1) = \sum_{j=1}^{N} \frac{F_{ij}}{F1} \times (F_{ij} - M_i)^2 \quad [2]$$

where: N is the total number of documents in the file
F_{ij} is the number of times a word (C_i) appears in the document (D_j);
F_i is the number of times C_i appears in N;
M_i is the mean value of C_i in D_j

This variance measure selects words with high, middle or low frequencies in the total file if they appear often in a restricted number of documents. Their ratio *number of occurrences / number of documents* is consequently high which means that words which appear frequently in the file but which are distributed over a large number of documents are not considered relevant for building the key-word list of a STAS system. This result is coherent with the goal sought: an early warning system should be capable of identifying elements in an information flow which are repeated in a limited organizational context. We have shown elsewhere that a high value of the above ratio can be explained by the publications of a limited and cognitively coherent group of actors [8].

d.) The last step of the LEXINET process is a manual validation of the machine constructed key-word list. Dedicated software was developed and tested in the Chemco experience in order to facilitate this validation exercise. The work was done on a MacIntosh using hypercard. The result is called CANDIDE [9]. Our goal in developing this software was to create a micro-computer workstation for managing STAS key-word dictionaries. We will give examples to illustrate how the CANDIDE workstation was used in the Chemco study later in this paper.

3.2) Dynamically updating surveillance techniques

The day-to-day activity of a Chemco department results in a constant flow of new documents. These documents constitute a source of information about changes in the department's centres of interests. LEXINET enables an Information Service to periodically open the department's electronic folder in order to:

- build a STAS key-word list of these interests at a given moment in time ('t');
- compare the list generated at 't+1' with the one existing at 't'.

The goal of this comparison is to update the STAS list in order to keep in step with the scientific and technological reality of the Department. New words focus attention on changing centres of interest. They suggest the need of orienting the department's view of its external environment differently, of modifying in other words its surveillance techniques. This need is precisely what bottom-up planning exercises are designed to detect. Search strategies for downloading information from external host services should be constantly modified in order to target emerging concepts.

The flow modeling methods which will be described shortly are designed to provide an Information Service with arguments to convince a department of the need to look at its world differently. However, a department is not a monolithic block. Its composed of services, working groups, project committees and individuals with specific interests and objectives. The question was raised earlier about how to locate the people in the network that should be alerted to changes in the competitive environment of the company. An Information Service has to get the right information to the right place at the right time. We will now discuss an algorithm we are working on in order to do so.

4) Getting the right information to the right place at the right time

The STAS key-word list is a lexicon defining a department's scientific and technological interests at a particular moment in time. However, it is an undifferentiated word list and consequently of little use as a dispatching device. An Information Service has to know who is doing what in the firm if it is to efficiently organize its early warning alert system. This means that the STAS dictionary has to be partitioned. The goal is to correctly describe the activity underway in the different functional units of a department.

In order to achieve this goal we are working on a three step programme. The first is to develop an appropriate model of Chemco's organizational structure. The second is to automatically distribute the words of the STAS dictionary over the different classes of the model. The rules adopted for assigning words to classes should result in a clear definition of a functional unit's field of respons i bility in the Chemco structure. This means that an isomorphous matching must be sought between the word list defining the extension of a class in the model and the one needed to describe a unit's organizational prerogatives. If this condition is met, the Information Service will be able to dispatch its early warning messages to the right place in the Chemco network for evaluation and action. The third part of our programme aims at developing an algorithm for carrying out this dispatching automatically.

4.1) Modeling an information structure

We decided to represent the organizational structure of Chemco by means of a three layer classification scheme. The department is at the root of a tree; the sectors in each department are located at branches leading to the leaves of the tree; these leaves are the working groups of the organization which can be composed of individuals located in different branches of activity. At 't' the organizational charts of Chemco list the members of each department and give their position in its hierarchical structure. Over time we expect that the units themselves will change. Old Departments will be replaced by new ones, the sectorization of the firm will evolve, the working group configurations will change rapidly given the growing concern to encourage flexibile responses to environmental change. However, the three-level hierarchical structure should itself remain a satisfactory model of Chemco's in-house organization.

The name of the author of a document is used to automatically distribute the documents in an electronic folder over the different classes of the three layer classification scheme just described. A document can be classed as belonging to two or more sectors if it is co-authored by the members of a multi-sectorial working group. While individuals can belong to several different working groups, in the organizational charts of Chemco they will appear as a member of only one department and as working in only one sector. In other words, working group membership is an indication of intersectorial cooperation. This explains why several classification codes can be automatically assigned to the same document.

The coding itself is carried out in two steps. A first pass serves to locate the functional unit(s) of a department in which the author(s) of a document is (are) located; a second pass then controls the source of the document and assigns it to either the formal or informal information categories discussed above. New words appearing in informal sources are weighted as being more important than those extracted from formal information sources. Our hypothesis is to consider informal communications (notes, letters, reports . . .) as being more important for defining new resource allocation strategies than the content of patents and articles which are written on the basis of research whose importance has already been recognized. For the moment, however, the usefulness of pursuing this kind of reasoning has not yet been tested. It might be that the distinction is unnecessary, but we have not yet progressed sufficiently in our research to answer this question.

When documents of a department are coded, the codes are added to the list of key-words already assigned to these documents as a result of the LEXINET procedures. It is consequently possible to automatically generate a file containing the association values of each word/code pair. The value is computed on the basis of the co-presence of words and codes in the same document. The association coefficient A_{ij} described in equation 1 above was used for this computation. In this case, A_{ij} measures the probability of finding a given organisational code 'i' in the same document as the word 'j' and, inversely, the probability of having 'j' if 'i', given both the number of times 'i' and 'j' appear together in a document (C_{ij}) and their respective frequencies in the total file (C_i and C_j).

This word/code file is used to verbalize the content of work carried out in each functional unit of a department. The rules used for this verbalization are designed to transform the STAS dictionary into a dispatching device for the Information Service's early warning system.

4.2) Defining responsibilities in an organizational structure

The words which are automatically assigned to the different classes of our organizational model should constitute a list of appropriate descriptive elements delimiting the scope of a functional unit's field of activity in the Chemco organization. We are currently experimenting the use of the following rules in order to achieve this goal.

Given that the organizational structure of Chemco is considered as a tree having the various departments as its roots and the working groups as its leaves:

- a word associated with several nodes appearing in the same branch of the tree will be assigned only to the node for which its association value A_{ij} is the highest. Multi-assignments of words to nodes do not exist.
- words are assigned to one and only one node in the tree in order to describe the subjects which are of specific interest to the functional unit located at that node;

- a functional unit has direct responsibility over the nodes situated below it in the classification scheme. This means, for example, that a working group's word list can be directly added to the one associated with its sector node in the hierarchical structure, but that the opposite is not true. The scope of sectoral responsibility is wider than the scope of a working group's area of interest. The words which are specific to a sector node define the content of this wider range of responsibility.

These rules can easily be understood given our previous discussion of an Information Service's need 'to find a spokesman'. In order to send the right information to the right place in the network a clear distinction has to be made between the content of the work being carried out at the group, sector and departmental levels. This can be done by ensuring that word lists corresponding to the activity of each functional unit are distinct from one level to the next. However, while it is probably correct to assume that people working at the group level are often the most technically competent to evaluate the meaning of changes in a firm's competitive environment, only sector or department heads are generally able to effectively engage collective action that might lead to the adoption of new resource allocation strategies. An Information Service cannot ignore the realities of organizational power in carrying out its bottom-up planning activities. If information is dispatched to a working group, a copy of it should also be sent to both sector and departmental heads in order to avoid unnecessary conflicts over prerogatives.

4.3) Dispatching information in a distributed network

We have not yet completed our study of the results obtained after automatically distributing the words of a STAS dictionary over the functional units of a Chemco department. Given that the sorting is done automatically, we need to organize quality control checks in order to avoid possible dispatching errors. We are working on this problem [10]. Tests have also begun on the dispatching algorithm itself. However, this work is at an early stage and we will consequently give only a brief account here of our hypothesis.

The structured STAS dictionary provides the Information Service with a benchmark for measuring what is important and what is not in the firm's environment. This environment can be monitored in a variety of ways. For example, the Chemco Information Service periodically downloads information from external host services using the same query formulation to up-date in-house files. It systematically surveys the patents and scientific publications of its most immediate competitors. It intends to use the new lexicon management techniques of LEXINET to focus attention on emerging concepts of possible interest to the firm. All in all the incoming flow of information can be quite substantial. About 5,000 downloaded documents a year serve to track changes in the firm's technological environment at the present time. However, Chemco officials expect this figure will increase rapidly in the next few years.

We can represent the impact of this external information flow on the internal in-house knowledge base (ie. the structured STAS dictionary) in the following way. Each functional unit (I) is characterized by a vector of word associations (I_k^t) where I_k^t is the association value between a word 'k' and the code 'I' at time 't'. Each new downloaded corpus represents the impact of the external world on the local in-house environment. This impact can be measured by assuming that it will lead to a new definition of a functional unit's scope of interests (I_k^{t+1}). Since both vectors (I_k^t) and (I_k^{t+1}) are defined on the same vocabulary, we can compute the association links on the two sets I^t and I^{t+1} by defining their vectorial cooccurence as follows [11]:

$$C_{I^t I^{t+1}} = \sum_k I_k^t \times I_k^{t+1} \quad [3]$$

This vectorial cooccurrence definition enables us to use the association coefficient defined in equation 1 above to measure the cosine of the directional change in a functional unit's interest when a new corpus is added to an existing knowledge base. It can be mathematically shown that if the downloaded corpus lies completely outside a unit's field of interest or, on the contrary, is completely isomorphous with that field of interest, then no directional change will be observed (ie. the cosine value will be high, about 1). A strong directional change, which is measured by a low cosine value, means that the impact of the external world on a functional unit's working environment could possibly be quite strong. Consequently, the Information Service would want to dispatch its early warning messages to those functional units which have the lowest cosine values.

5.) Modeling the information flow

Dispatching information to the right place in the network is not enough to ensure that action will follow. Management must be convinced of the need to redefine resource allocation strategies given changes in a firm's competitive environment. An Information Service can not expect management to take the time to read a flow of documents in order to evaluate the importance of these changes. More synthetic measures are required to win the support of a spokesman in management circles. We have been attempting to develop these measures in a programme aimed at mapping the dynamics of science and technology [12]. We will now give an example of our results.

5.1) Problem area maps

Chemco is a multinational firm with activities in the gas industry. A product department carries out research into the chemical, physical and

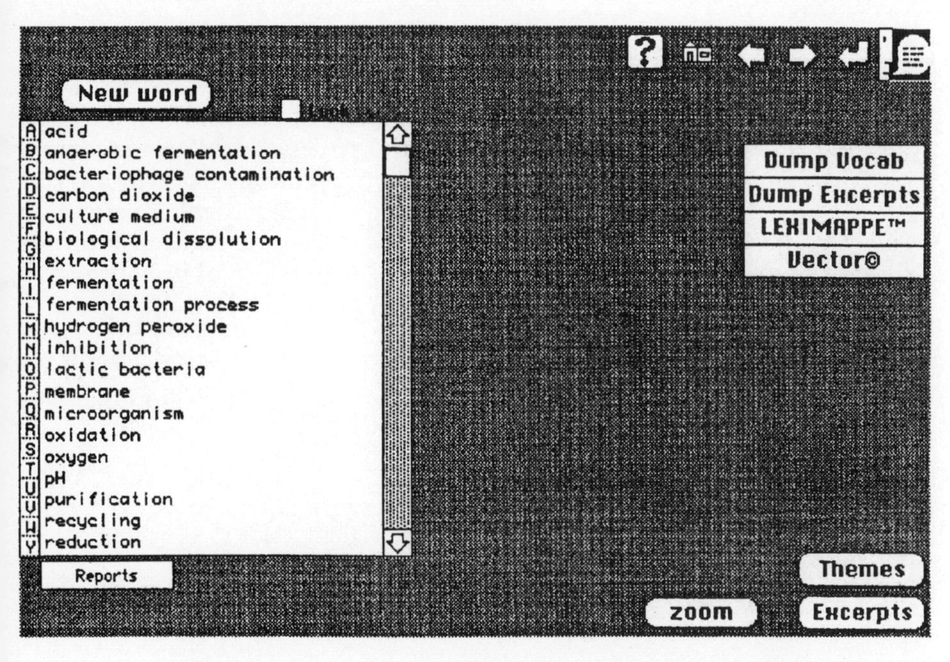

Figure 1: A word list defining the activity of a Chemco working group on fermentation processes

biological uses of gas products. In this last sector, a working group was set up to study fermentation processes. Figure 1 gives a print-out of a CANDIDE window showing the list of key-words automatically assigned to the code of this working group's slot in the firm's organizational structure. These words were extracted from the department's electronic folder using the different LEXINET procedures described above.

Figure 2 presents another CANDIDE window. The information visualized is a map of one of the word association structures found in the five patents that the working group took out over the eight years of its activity. This map suggests a *problem* area that was studied by the working group: that of inhibiting the growth of microorganisms (lactic bacteria) on culture mediums. A problem area map is a way of packaging information for evaluation by management. We have shown elsewhere that when a new corpus is added to an existing knowledge base, the core of a problem area map will remain intact (ie. the dark-lined loops) if there is a relatively homogeneous increase in the basic statistics (C_i, C_j and C_{ij}) used to compute the A_{ij} coefficient defined in equation 1 above [13]. Figure 3 shows this to be the case in our example. The new map contains much more information than the one shown in Figure 2. The differences can be traced to the impact of 17 patents taken out by one of Chemco's immediate competitors in the gas industry. We used the equation 3 above to identify the fermentation unit as being the one the most likely to be

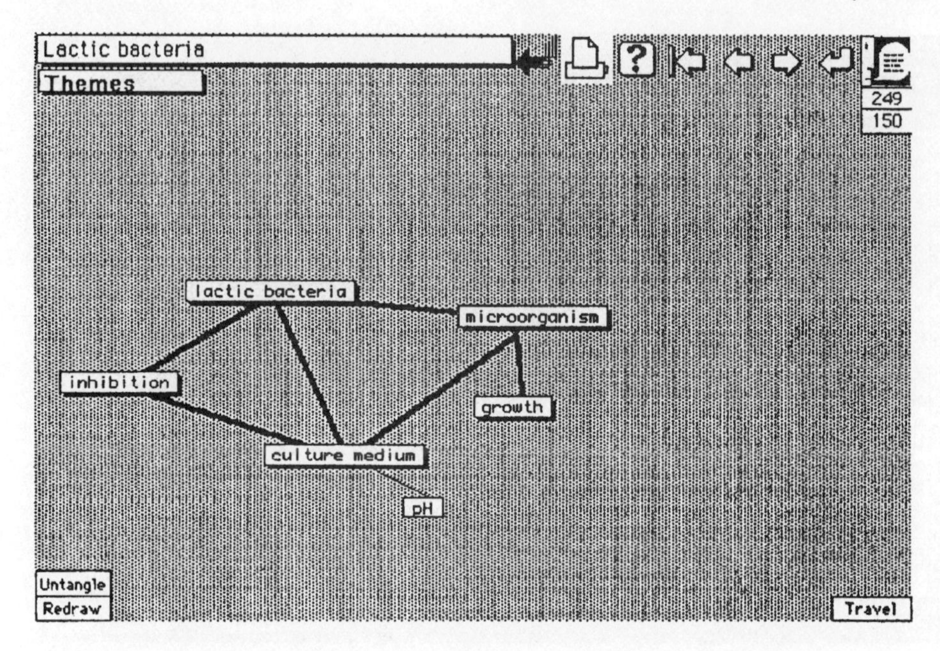

Figure 2: A map defining a problem area of particular interest to Chemco's working group on fermentation

concerned by the content of these downloaded patents. We then mapped the association structure of the document set obtained when the 5 Chemco patents were combined with the 17 competitor patents. We see that the competitor company is working on exactly the same type of problem as Chemco, but seems to be considering a wider range of subjects in connection with it. The light-lines on Figure 3 indicate access routes to other problem area maps that can be dynamically explored using the CANDIDE workstation as a means of 'navigating' in the cognitive space of the fermentation research network [14].

The map on Figure 3 is an early warning device. However, it has once again been generated automatically. Quality control tests should therefore be carried out before the Information Service alerts the Chemco working group on fermentation as to what is being done by its competitor. The CANDIDE workstation can be used for these quality control tests. It allows a user to test different document retrieval strategies suggested by the word association patterns on the problem area map and to modify the STAS dictionary by manually introducing new key words to improve the precision of those which appear on the maps.

5.2) From maps to document retrieval

For quality control tests, it is important to note the existence of word loops and word chains on a problem area map. Word loops indicate the

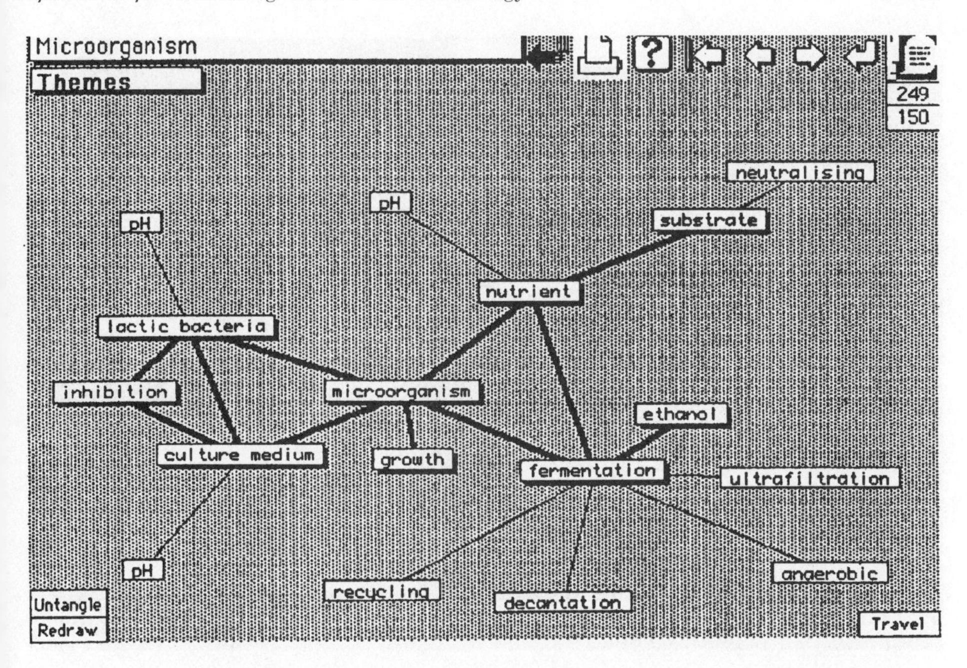

Figure 3: A map illustrating the impact of downloaded information on an 'in-house' definition of a problem area

existence of coherent problem areas whereas word chains suggest ways of linking these problem areas together. This observation is important for document retrieval purposes. Words in a loop should be linked together with an 'or' operator because their individual document sets will tend to overlap to a large degree. The 'or' can consequently be used to avoid silence especially since the danger of noise is not high. However, when it comes to using word chains as the starting point for query formulation the 'and' operator should be applied. The reason for this recommendation stems from the fact that there is not necessarily a significant overlap between the document sets corresponding to a word chain. Noise is therefore a more difficult problem to manage in this case than it is in the former.

Figure 4 shows a new CANDIDE window which is used for elaborating document retrieval strategies before launching them on-line for questioning external host services. 'Nutrient' and 'substrate' are connected in this Figure by the boolean operator 'and' ('x' in the syntax of CANDIDE). The two terms are 'chained' together on the association map shown in Figure 3. The Information Service of Chemco suspected that these two words should in fact be considered as a compound term. By reading the patents in which these two terms appear together it is easy to improve the STAS dictionary. The three patents to be consulted are identified in the window 'qui en parle?' on Figure 4. By clicking on the

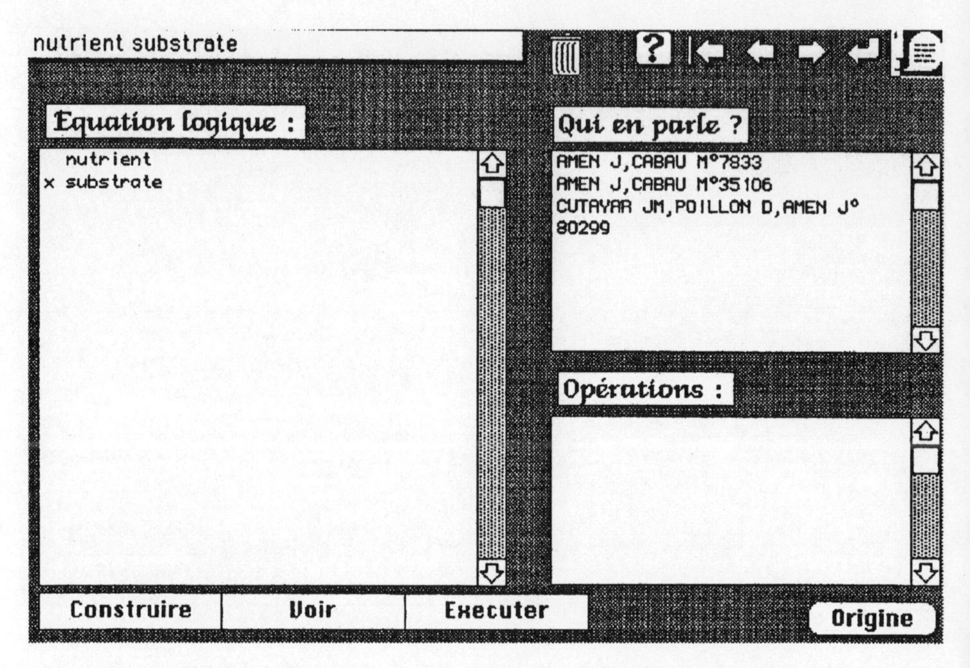

Figure 4: A CANDIDE workstation technique for testing document retrieval strategies 'in-house' before questioning external host services

Cutayer, Poillon and Amen patent, CANDIDE will take us to the document itself. The title (TI) and abstract (AB) of the patent are located in the window 'index the word' on Figure 5. The document can be browsed and when new words are required for indexing, they can be added to the list of indexing terms automatically generated by LEXINET. An example is given in Figure 5. After reading the patent, we decided to replace the uniterms 'nutrient' and 'substrate' by the compound term 'nutrient substrate'. In order to carry out these operations we simply 'unlocked' the key words listed in the window on the left of the screen. This action enabled us to correct the list as required. CANDIDE will automatically add the new compound word to the STAS dictionary.

Conclusions

This paper has reported on work underway with a large chemical producer in France, Chemco. It was organized in three parts.

The first part of the paper discussed the problem of managing information flows in a large multinational company. The conclusion of this section was that we are in the midst of a paradigm change. 'Just-in-time' production theories are making their way into the information practices of large firms. The question raised is that of knowing how to get the right information to the right place at the right time.

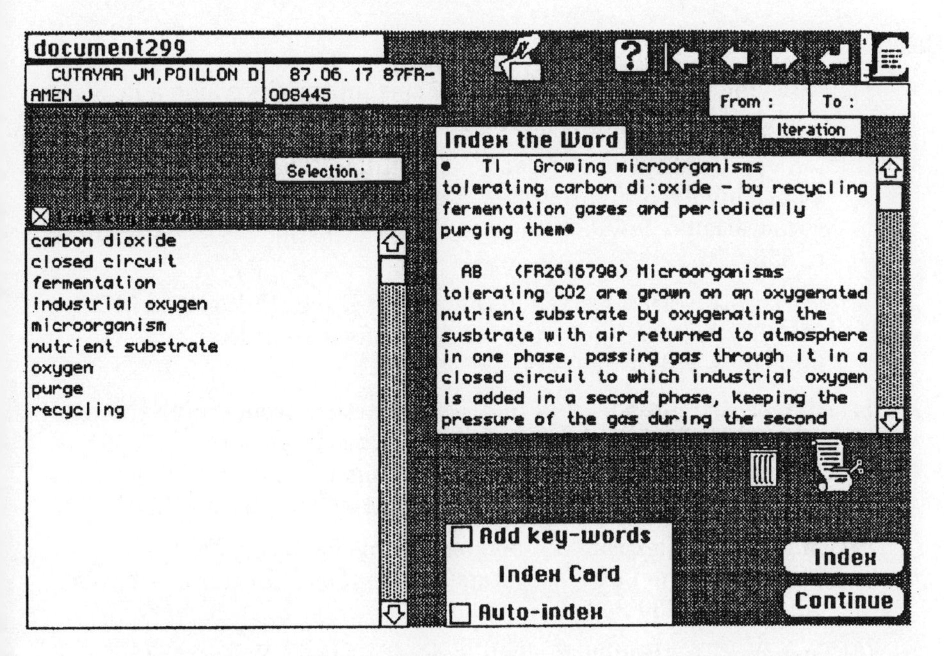

Figure 5: An example of a quality-control test aimed at determining the information value of a problem area map

The second part of the paper examined the impact of this paradigm change on information management practices. Existing tools are basically designed to access stocks of stored information. New tools are needed to manage information flows in order to meet the requirements of a 'just-in-time' policy. After describing these requirements, we introduced the general notion of infometrics as a way of meeting them. Infometrics designates work aimed at defining benchmarks for quality control tests in an information flow management context.

The third section served to give examples of work currently underway to consolidate an infometric approach to the mapping and measuring of science and technology. The Information Service of Chemco would like to initiate bottom-up planning exercises. This means systematically monitoring external information sources in order to detect changes in the competitive environment of the firm which might affect its level of scientific and technological competence. This paper discusses ways of building a scientific and technological alert system.

References

[1] P. Levine and J-C. Pomerol, Systèmes interactifs d'aide à la décision et systèmes experts, Herms, Paris, 1990.

[2] C-F von Braun, H.G. Fischer, A.E. Müller, 'The need for and the issues involved in integrated R and D planning in large corporations', *Int. J. Technology Management*, Vol. 5, No. 5, 1990, p. 559-576.

[3] Work in this area started in the mid 1960s: G. Salton and M. McGill, 'An introduction to Modern Information Retrieval', MacGraw-Hill, New York, 1983.

[4] An interested reader will find several articles concerning this general problem in the Conference proceedings of the RIAO91 Meeting at Barcelona. RIAO91, 'Intelligent Text and Image Handling', Conference Proceedings, Barcelona, Spain, April 1991.

[5] Chartron, G 'Lexicon Management Tools for Large Textual Databases: the Lexinet System', *Journal of Information Science*, **15**, 1989, p. 339-344.

[6] Turner, W.A., Buffet, P., Laville, F., 'LEXITRAN: For an Easier Access to Patent Databases', *World Patent Information*, Vol 13, No. 2, 1991, p. 81-90.

[7] B. Michelet, 'L'analyse des Associations' Thèse de Doctorat, DEA Information Scientifique et Technique, Université de Paris VII, 1988.

[8] M. Callon, J. Law, A. Rip, 'Mapping the Dynamics of Science and Technology', MacMillan, London, 1986.

[9] The CANDIDE programs were written by B. Michelet. Their use is not limited to the application discussed in this paper. They have been tested in the 'reading workstation project' which was undertaken in order to explore the general problem of developing electronic aids for social science research by the steering Committee of the new French National Library. See the article by D. Chouchan, 'Du livre à l'ordinateur', *La Recherche*, **228**, Janvier 1991, Vol. 22, pp. 96-100. A similar application has been developed by G. Teil, 'Une sociologie assistée par ordinateur', Thèse d'Universit, CSI/Ecole des Mines, 1991.

[10] A micro-computer programme designed to allow interactive evaluation of word/code lists was developed. See M. Kaltenbach, W.A. Turner, F. Laville, 'LEXITRAN-Mediated access to Patent Databases' in *Journal of Information Science,* **17**, 1991, p. 13-20.

[11] B. Michelet, L'analyse des associations, Thèse du Doctorat, Université de Paris VII, 1988, p. 78.

[12] W.A. Turner, G. Chartron, F. Laville and B. Michelet, 'Packaging information for peer review: new co-word analysis techniques' in A.F.J. van Raan, 'Handbook of Quantitative Studies of Science and Technology', Elsevier, North Holland, 1988.

[13] W.A. Turner, B. Michelet, J.P. Courtial, 'Scientific and Technological Information Banks for the Network Management of Research', *Research Policy*, Vol. 19, No. 5, Octobre 1990.

[14] F. Laville et W.A. Turner, 'Guide Pratique pour l'Utilisateur de Candide', Société Transvalor, 60 bd. St. Michel, 75006 Paris, 1990.

The information environment and the productivity of research

Michael E.D. Koenig

Graduate School of Library and Information Science, Rosary College, River Forest, Illinois, USA

Abstract:

This study attempts to examine the relationship between the R&D productivity of major U.S. pharmaceutical companies and the information environment of those companies. R&D productivity is defined as new drugs approved per dollar of research budget. This measure is further refined by weighting it in regard to 1) whether or not the FDA regards the drug as an important therapeutic advance 2) the drug's chemical novelty, and 3) the filing company's patent position with regard to the drug, an indicator of where the bulk of the research was done. Productivity among large pharmaceutical companies, as measured by this metric, differs quite dramatically. Data on the information environment of the companies was gathered from questionnaires distributed to individual researchers at those companies. Productive pharmaceutical R&D information environments were found to be characterized by:

- *Less company concern about the proprietariness and confidentiality of company data and*
- *A greater frequency of use of company libraries and information services.*
- *Greater openness to outside information, including greater attendance at external professional meetings;*
- *Researchers report a greater proportion of their information seeking behavior is directed toward browsing and staying abreast;*
- *Greater organizational encouragement of the use of information services, and of information seeking beyond one's immediate needs.*
- *A work environment that is perceived to be more egalitarian and where status indicators are less obvious.*

Introduction

The study reported here examines the relationship between the productivity of new drug development across a number of pharmaceutical companies versus the information environment of those companies as described by researchers in those companies.

Productivity is a concern never out of vogue and at the moment, coupled with international competitiveness, it seems to be particularly in vogue. The productivity of the research process is a particularly interesting aspect, and one that has been relatively little addressed. A principal reason for this lack of attention has been the difficulty in quantifying research productivity in a replicable and meaningful fashion.

Organizations establish information services to make themselves and their individual members or constituents more productive, whether pharmaceutical researchers, or faculty, or students, or the body public. Yet, we know very little about the relationship of the services we provide and the information environments we create, with the productivity of the organizations and individuals we support (BLAGDEN 1980, KOENIG 1990). The principal problem of elucidating that relationship is that it is very difficult to measure productivity, particularly the productivity of the persons or functions supported by libraries and information centers. The people who are supported by libraries and information centers are knowledge workers, and it is notoriously difficult to measure the productivity of knowledge workers. See Koenig 1990 for a recent and fairly thorough review of this topic.

The pharmaceutical industry is perhaps unique in the environment which it provides for studying the subject of research productivity. Foremost is the fact that the pharmaceutical industry is clearly very research dependent. Indeed, Schwartzman describes the pharmaceutical industry as being characterized by "competition by innovation" (Schwartzman, 1976). The phrase often heard within the industry is "the NDA (New Drug Approval) is the name of the game." Additionally, the products of that research effort, new pharmaceutical agents, are discrete, enumerable, reasonably commensurable, and the data is publicly accessible. The rigor (and considerable expense) of the Food and Drug Administration (FDA) new drug approval (NDA) process ensures that therapeutic agents are not submitted for approval lightly, and that therefore in using NDAS as the basis for our output metric we are counting only non-trivial outputs. Furthermore, the granting of the NDA is open and relatively bias free, and is as unlikely to be subject to halo effect[1] as one is apt to find in this imperfect world. Finally, the industry is relatively unconcentrated, so that there is a sufficient number of organizations pursuing the same goals, some two dozen major firms in the U.S., to expect to find some degree of statistical regularity.

The productivity metric is compared against the information environment. The information environment has been defined by Taylor as follows: "An organizational information environment is made up of those variables (a) that affect the movement of information messages into, within, and out of any definable organizations; and (b) that determine the criterion by which the value of information messages will be judged in that context" (Taylor, 1982).

The information environment then, as defined here, consists of those mechanisms, both formal and information, that we provide, whether consciously or not, for the accessing and dissemination of information and knowledge, and the creation of an environment that encourages and facilitates the use of those mechanisms. This information environment encompasses formal mechanisms such as libraries and database searching, semi-informal mechanisms such as professional meeting attendance, and informal mechanisms such as lunchtime conversation, and the structures or climates set up consciously or unconsciously to facilitate those mechanisms.

Methodology

Data Collection

Productivity

Central to this analysis is the measure of pharmaceutical research productivity, the main dependent variable in this study. The metric developed is an output/input measure based on the Food and Drug Administration (FDA) approved pharmaceutical agent as output, and R&D budget as input. This measure is refined in three ways. First, only those NDAs that are new chemical entities are considered. Medical devices, combination products, new uses of an existing compound and minor molecular modifications are excluded. The rationale for this distinction is that it reasonably reflects the distinction between NDAs that represent original research and those which are generally derivative.

Second, those drugs on which the company enjoys patent protection are weighted more heavily, on a ratio of 2 to 1, than those drugs without such

---

[1] The work of Anderson et al. (Anderson, 1978) indicates a substantial halo effect in expert judgement of research quality, and work by Koenig supports the same conclusion (Koenig, 1983B)

protection. The rationale for this is that if a pharmaceutical company has done the development work, but not the original research, it will typically not enjoy patent protection. In those cases, the patent may have been assigned to the company from another company whose patent was licensed, or the discovery may have originated externally and been reported in the open literature. The patent position then is an indicator of how much of the R&D process was undertaken by the company.

Third, the drugs are weighted by medical importance. This weighting is based on an annual internal FDA study which classifies new drugs as to therapeutic potential. Those drugs regarded by the FDA as being particularly important are classified as 'Important Therapeutic Gains.' Such drugs, ITGs, are weighted in this study in a ratio of 2 1/2 to 1 to non ITGs, this ratio being derived simply from the inverse of the numerical proportion of ITGs to non-ITGs.

These refinements and weightings resulted in a drug output score. The composite drug output score can be expressed as:

Score = (5 x I-P) + (2.5 x I-WP) = (2 x N-P) - (1 x N-WP), where

I-P = Important (ITG) Drugs with Patent Protection

I-WP = Important (ITG) Drugs without Patent Protection

N-P = Non ITG Drugs with Patent Protection

N-WP = Non ITG Drugs without Patent Protection

The NDA was evaluated for the years 1965 to 1983. The basic measure of pharmaceutical company research input was the R&D budget for the years 1965-1978.(The availability of such data was also one of the reasons for limiting the analysis to companies in the U.S.; financial reporting requirements in most other countries are typically much less rigorous, not requiring sectoral break outs by line of business.)

The basic sources for company R&D budget data are the companies' annual reports and the 10K reports filed with the Securities and Exchange Commission (for those companies publicly traded in the U.S.). Numerous complications remain, however, for example, R&D data has been required in 10K reports only since 1973, and even now it is not yet required that companies break down their R&D by product line. Furthermore, if a company, such as American Home Products, has functionally and organizationally separate subsidiaries, such as Wyeth and Ayerst[1], financial data need not be broken down by subsidiary.

---

[1] American Home Publications has recently reorganized, but for the period of this study, Wyeth and Ayerst were run as quite separate companies.

The data used here have been further modified and refined as a result of numerous phone calls and conversations with the author's former colleagues in the pharmaceutical industry, many of whom were gracious enough to volunteer information whose release was clearly in violation of company policy. Needless to say, these sources cannot be acknowledged. These sources were used to disambiguate situations such as the above example of Ayerst and Wyeth, and to extract the not relevant R&D costs, such as consumer products R&D (for companies such as Bristol Myers, that is very large component of the reported R&D costs) from the reported figures.

There is an extensive literature on 'drug lag' relating to the length of time between when the basic research is done and when the drug is approved (Lasagna 1980; Wardell & Scheck, 1983). This lag is generally regarded to be of the order of seven years. The lag here between expenditure of dollars and approval of the NDA is slightly less. In any case, both pharmaceutical company performance as defined here and their R&D expenditures tend to be quite stable over time (Koenig, 1983), so the analysis is unlikely to be affected in any significant fashion by the drug lag phenomenon.

The Information environment

The data concerning the information environment was gathered by means of a questionnaire directed to researchers in the target pharmaceutical companies. The pharmaceutical companies themselves are not particularly enthusiastic about cooperating with investigators or such projects. For a number of reasons they are very wary about investigators from the academic world. The researchers as individuals, however, regard themelves as members of the larger scientific community and are typically cooperative in responding to questionnaries, if the topic seems reasonable, the questionnaire well thought out, and most importantly if the topic and the specifics of the questionnaire do not intrude into areas of proprietary concern and competitive sensitivity. The questionnaire used is included as Appendix A. The name and addresses of individual researchers were located in various professional society directories. The response rate to the questionnaire was 47%, a rather high rate of return for a questionnaire of considerable size and detail. The author attributes this high response rate to the relevance of the topic and to the fact that the questionnaire was, in the words of one respondent, "obviously put together by someone with an insider's knowledge of the pharmaceutical industry." The author would also like to acknowledge the support of Nancy Roderer of the Columbia University Health Science Library who supplied an extensive collection of questionnaires used in previous information use studies. These questionnaires provided numerous suggestions to the investigator.

Analysis

The principal technique used was multiple stepwise regression using the SPSS (Statistical Package for the Social Sciences) software system. The independent variables were the productivity metric for the pharmaceutical companies. The dependent variables were the various data elements characterizing the information environment.

Findings

The central findings are but displayed in the stepwise regression tableau presented below. (Figure 1.)

STEPWISE REGRESSION ON "PRODUCTIVITY"

Step	Independent Variable	Adj. R Squared	Beta Weights							F
1	Confidentiality (-)	0.23	-.48	-.44	-.37	-.36	-.36	-.31	-.31	33.1
2	Visit Library	0.40		.43	.42	.40	.40	.35	.33	38.0
3	Status (-)	0.49			-.31	-.32	-.29	-.30	-.27	36.4
4	Library Staff	0.55				.24	.26	.22	.19	34.1
5	Micro Hours	0.59					.24	.25	.25	34.0
6	Lib. Encouragement	0.63						.19	.19	32.1
7	Prof. Activities	0.64							.14	29.3

Confidentiality = Rating of own company's emphasis on Confidentiality.
Visit Library = How often the Library/Info.center is Visited.
Status = Agreement with: "Status indicators are relatively unsubtle."
Library Staff = How do you rate your Library/Info. center Staff?
Micro Hours = Hours spent per week on a terminal or Microcomputer.
Lib. Encouragement = Agreement with: "I am Encouraged to spend time in the Library/Info. center."
Prof. Activities = Rating of own company's encouragement of Professional Activities.

Figure 1

The dependent (target) variable in this regression is the productivity of the pharmaceutical companies. The first variable to enter, and entising with a negative correlation, is 'Confidentiality', the researchers' ratings of their own company's emphasis on confidentiality. The adjusted R squared is not large, .23, but it is significant. The more productive companies, at least as perceived by their own researchers, do not place as high an emphasis on maintaining the proprietariness and confidentiality of company data and information as do the less productive companies.

The second variable to enter, and adding markedly to the adjusted R squared, is the number of times per week the researchers reported visiting the library. Researchers in the more productive companies clearly report visiting their company libraries and information centers more often than do the researchers in the less productive companies. Interestingly, this propensity to visit the library does not appear to be a function of proximity. The researchers were asked to report the time to travel to their library/information centers in minutes. The correlation between productivity and time to travel to the library was quite small (-.145). The researchers in the more productive companies did have a very slight tendency to report shorter travel times to their library/information center, but this correlation was very much less than that of the frequency of visiting the library (.467).

The third variable to enter, and negatively, is 'Status', agreement with the statement 'Status Indicators in this company are relatively unsubtle as compared to other companies'; that is, researchers in the more productive companies perceived their companies to be relatively more egalitarian,and in particular, the display of status indicators was perceived to be more subtle and less obvious than in the less productive companies.

The next variable to enter was 'library staff', the response to the question "How would you rate the technical competence of your library / information center staff?" This relationship was positive; the researchers at the more productive companies tended to rate their library / information center staff higher than did the researchers at the less productive companies.

Entering next, with increasingly smaller additions to the adjusted R square were:

Micro Hours, The Hours spent per week on a terminal or microcomputer; Library Encouragement, Agreement with the statement: "I am encouraged to spend time in the library / info center;" and Professional Activities, The researchers' ratings of their own company's encouragement of professional activities.

The principal correlations found between company productivity and aspects of the information environment as reported by the researchers in those companies were:

Confidentiality	-.483** Rating of own company's emphasis on confidentiality.
Visit Library	.467** How often the library/info center is visited.
Beyond	.413** Agreement with: "Company encourages information seeking beyond immediate need."
Status	-.410** Agreement with: "Status indicators are relatively unsubtle."
Lib. Encouragement	.398** Agreement with: "I am encouraged to spend time in the library/info center."

Prof. Activities	.382** Rating of own company's encouragement of professional activities.
Prof. Meeting	.381** Days per year attending professional meetings.
Other Organizations	.287* Agreement with "I am encouraged to seek work related information from colleagues in other organizations."
Browse	.286* Prcportion of library time used for browsing and keeping abreast rather than seeking specific information.
Library Staff	.283* How would you rule the technical competence of your library /info center staff?
Micro Hours	.280* Hours spent per week on a terminal or microcomputer.

(1 tailed significance * = .01, ** = .001)

The general conclusions that can be drawn from the data above appear to be that productive pharmaceutical R & D information environments are characterized by:

- Less company concern about the proprietariness and confidentiality of company data
- Greater openess to outside information, including greater perceived encouragement to seek work related information from colleagues in other organizations
- A greater frequency of use of company libraries and information services
- Researchers report a greater proportion of their information seeking behavior is directed toward browsing and staying abreast
- Greater organizational encouragement of the use of information services, and of information seeking beyond one's immediate needs
- A work environment that is perceived to be more egalitarian and where status indicates are less obvious.

In addition to the correlations of some significance found above, some of the non-statistically significant relationships are very interesting as well.

There was no significant correlation with the amount of time reported spent information seeking, either at work, or external to the job (-.05, and .03). This is consistent with previous findings by King *et al.*, 1984 that the amount of time spent in information seeking behavior is relatively stable across organizations of quite different type. It is also interesting that researchers in the more productive companies, while not spending more time overall in information seeking, do manage to find more time, that is, to allocate more of their information seeking time to browsing and staying abreast.

There was no significant correlation with either the number of presentations made, within the organization, or externally (-.17,and .02), or the number of presentations attended, internally within the organization or externally (-.02, and -.15). This is intriguing in contrast to the rather significant correlation with the number of days per year attending professional meetings (.381). These findings are certainly not inconsistent with the conventional wisdom that the utility of a professional conference is not so much in the formal presentations, but in the informal peer to peer communications.

The Gatekeeper Phenomenon

The study did not attempt a sociometric analysis of interpersonal communication within the individual companies, so it does not directly shed light on the existence or functioning of the gatekeeper phenomenon observed and elucidated by T. J. Allen, and examined in several subsequent studies (Allen 1970, 1971). The questionnaire did, however, ask whether the respondents did perceive their own company's R & D management as being particularly aware of and responsive to the gatekeeper phenomenon. The investigations would have hypothesized that the more productive companies would have been more characterized by such an awareness, but no significant correlation was found (-.08). It is perhaps noteworthy that almost no one (one case) did not answer this question, nor did any one query what was meant by 'gatekeeper.' At least within the pharmaceutical R & D community, the concept of gatekeeper appears to have become standard terminology. The researchers did report themselves as being more consciously aware of who the gatekeepers in their organization were, than they perceived their managers to be (a score of 3.4 versus 2.9 on a 1 to 5 scale where the larger number indicates greater perceived awareness).

Discussion

While neither the correlation coefficients nor the adjusted R2 between the target (dependent) variable and the independent variables are dramatically high, they appear in context however to be very significant. The R&D process is extremely complex with many richly interwoven components, such as management style, physical resources, extent of laboratory support services, not to mention, of course, the competence and creativity of the researchers themselves and the various factors that are likely to draw good researchers to a particular company, reputation, management style, location, compensation, plus of course, corporate culture and information environment, at which point we have woven ourselves back to the starting point. In such a complex environment, the strength of the relationships found are indeed provocative and revealing.

By way of contrast, in a previous study the two conventional indicators of research and scholarly merit, citation analysis and expert opinion, were compared with the same metric of productivity as used here (the only difference being that the data was from a slightly earlier time period), and the relationships found were not as strong as those found in this study (Koenig, 1983 A&B). This, despite the fact that all three components of that study, 'drugs per dollar,' citation data, and expert opinion were to a large degree simply three different measures of the same phenomenon, the quality of research.

What that implies is that if one assumes that generating the maximum number of good new drugs per dollar of research budget is what pharmaceutical research is about, that is if one assumes that the metric of pharmaceutical R&D productivity used here is in fact close to a true metric, not merely an indicator, then one could in principle predict pharmaceutical research productivity better by measuring a few key indicators of the information environment than one could by conducting a conventional citation analysis or by soliciting expert opinion. This result is surprising and provocative.

A Caveat

One must be careful in interpreting these findings. As is so often repeated, correlation is not causality. Differentials in research productivity and differentials in the research environment may both be a function of some third factor. It might for example be the case that some companies manage to hire more brilliant researchers, and that more brilliant researchers produce more successful research and that they also demand or create a different information environment. Or, it might be that causality flows not from greater use of the library/information center or from greater attendance at professional meetings to research productivity, but rather that success in research creates a climate which permits a company to encourage both greater use of the library/information center and professional meeting attendance. Nonetheless, the relationships are highly suggestive, and help illuminate a vital area about which we know far too little.

References

ALLEN, THOMAS J. 1970. 'Roles in Techical Communications Networks.' In: Nelson, Carnot E.; Pollock, Donald K., eds. 'Communication Among Scientists and Engineers'. Lexington, MA: Heath Lexington Books; 1970. 191-208. LC: 71-129156.

ALLEN, THOMAS J. 1977. 'Managing the Flow of Technology: Technology Transfer and the Dissemination of Technological Information Within the R&D Organization'. Cambridge, MS: MIT Press; 1977. 320p. ISBN: 0-262-01048-8; LC: 76-57670.

ANDERSON, R. C.; MARIN, F.; MC ALLISTER, P. 'Publications Ratings Versus Peer Ratings of Universities,' *Journal of the American Society for Information Science*, **29**(3): 91-103, May 1978.

BLAGDEN, JOHN F. 'Libraries and Corporate Performance: The Elusive Connection.' In: Taylor, Peter J., ed. New Trends in Documentation and Information: Proceedings of the Federation Proceedings of the Federation Internationale de Documentation (FID) 39th Congress; 1978 September 25-28; Edinburgh, Scotland. London, England: ASLIB for FID, 1980, pp. 379-382. ISBN: 0-85142-128-8; LC: 81-1085618.

KING, DONALD W.; GRIFFITHS, JOSE-MARIE; SWEET, ELLEN A.; WIEDERKEHR, ROBERT R. V. 1984. 'A Study of the Value of Information and the Effect on Value of Intermediary Organizations, Timeliness of Services and Products, and Comprehensiveness of the EDB'. Rockville, MD; King Research, Inc.; 1984. 3 volumes in 1. (Submitted to the U.S. Department of Energy, Office of Scientific and Technical Information). NTIS: DE85003670; OCLC: 11712088; DOE: NBM-1078. Available, by permission, from: King Research, Inc., P. O. Box 572, Oak Ridge, TN 37831.

KOENIG, MICHAEL E.D. 1983a. 'A Bibliometric Analysis of Pharmaceutical Research.' *Research Policy*. **12**(1): 15-36, 1983 February. ISSN: 0048-7333. 7333.

KOENIG, MICHAEL E.D. 1983b. 'Bibliometric Indicators Versus Expert Opinion in Assessing Research Performance.' *Journal of the American Society Information Science*. **34**(2): 136-145, 1983 March. ISSN: 0002-8231.

KOENIG, MICHAEL E.D. 'Information Services and Downstream Productivity.' In: The Annual Review of Information Science and Technology. Martha E. Williams, ed. Vol. 25, New York: Elsevier, 1990, Chapter 2, pp. 55-86.

LASAGNA, LOUIS, Ed. Controversies in Therapeutics. Philadelphia: W. B. Saunders, 1980.

TAYLOR, ROBERT S. 'Organizational Information Environments.' In: Information and the Transformation of Society, G. P. Sweeny, Editor, Amsterdam: North Holland Publishing Co., 1982, pp. 309-322.

WARDELL, WILLIAM M. & SHECK, LORRAINE E. 'Is Pharmaceutical Innovation Declining?: Interpreting Measures of Pharmaceutical Innovation and Regulatory Impact in the USA, 1950-1980.' *Rational Drug Therapy*, **17**(1): 1-7, January 1983.

Biosequence searching and Its application in the pharmaceutical industry

Huei-Nin Liu-Johnson and James F. Corning

Chemical Abstracts Service, Columbus, Ohio, USA

I. Introduction

At a recent meeting of the Association of Biotechnology Companies (ABC) in Washington, D.C., outgoing ABC president Martin Nash remarked that the biotech industry finally "broke out of the lab" in 1990 to have a significant impact on human health. Nash cited two important milestones as evidence of this: revenues of biotech companies exceeded a billion dollars ($1.2 billion) for the first time, and over 100 million people were treated with the products of biotechnology. These products ultimately are based on two important classes of compounds, nucleic acids and proteins.

Nucleic acids (in the form of ribo- or deoxyribonucleotide polymers) constitute the heredity material, or genome, of living organisms. The order, or sequence, of the nucleotide bases [adenine (A), cytosine (C), guanine (G), and thymine (T) or uracil (U)] in the polymer determines the information content of the genome.

Similarly, proteins are polymers of amino acids. The sequence of amino acids in a naturally occurring protein is determined by the sequence of nucleotides in the gene that codes for that protein. Proteins are the most abundant organic molecules in cells and are important in all aspects of cell structure and function. Hormones, enzymes, blood factors, interferons, and antibodies are all examples of proteins. Some of these proteins have been genetically engineered and approved by the FDA for use as pharmaceuticals. They include insulin and growth hormone for treatment of growth and nutritional disorders, alpha and gamma interferons for cancer therapy, tissue plasminogen activator for treatment of cardiovascular disorders, erythropoietin for anemia and AIDS therapy, hepatitis B and haemophilus B vaccines, and a therapeutic monoclonal antibody (product report by Thayer, 1991). The development and marketing of other similar products are expected to continue in the coming years as the field of biotechnology continues to advance.

Developments in computer technology, molecular genetics research, powerful yet easily implemented recombinant DNA techniques, and automated sequencing methods during the last decade have not only fundamentally changed the way much research in the life and medical sciences is conducted but have also brought us to the point where projects such as the Human Genome Initiative for mapping and sequencing of the entire human genome are no longer dreams but reality (Human Genome Program Report, 1990). An area of pharmaceutical research that is gaining importance and advancing rapidly is the development of antibodies and antigens for immunotherapy. Monoclonal antibody-mediated therapy has been revolutionized by advances such as the definition of cell-surface structures on abnormal cells as targets for effective antibody action; genetic engineering of hybrid or modified monoclonal antibodies that are less immunogenic and more effective; and arming of such antibodies with toxins or radionuclides to enhance their effector function (reviewed by Waldmann, 1991). In addition, use of the epitope library approach (Scott, et al., 1990), which can easily and speedily test millions of peptides to derive discontinuous antigenic epitopes, finds application in the investigation of antibody specificities and discovery of mimetic drug candidates.

Not surprisingly, this growth in science and technology has led to a corresponding increase in the information base. This accumulation of information has occurred in printed materials such as journals, patents, and government regulations as well as in a number of both general and specialized bibliographic and substance databases. The pace of discovery and the appearance of the information have in many cases increased dramatically in the last 10 to 15 years; and, as a result of rapid developments in biotechnological methods and projects like the Human Genome Initiative, this increase in information is expected to continue. Besides the traditional information sources, databases containing nucleic acid and protein sequence information, especially from the patent literature, are of growing importance to the pharmaceutical industry.

Nucleic acid files such as Genbank and EMBL and protein files such as PIR and SwissProt are specialized databases that provide points of access to both biosequence information and bibliographic data and, when used in conjunction with any of the various biosequence analysis software packages, can be useful guides to predicting structure and function of a molecule based on its sequence data. Recently, Chemical Abstracts Service (CAS) has begun providing access to protein sequence and related bibliographic information via STN International (Mohindru *et al.*, 1990; Liu-Johnson *et al.*, 1990). The CAS Registry File contains some 170,000 protein and peptide sequences making it the largest such file of its kind and the only source containing both protein sequence data from journal and patent sources in the same file. Furthermore, the file is comprehensive in types of sequences. In addition to the naturally occurring wildtype and mutant sequences covered in PIR and SwissProt, the Registry File also contains sequence data for smaller peptides, synthetic proteins and peptides, and proteins and peptides that have been chemically or genetically modified. Thus, the CAS data can both complement and supplement the other protein sequence databases (Mohindru, Ambrose, and Corning, 1990).

In a competitive industry such as the pharmaceutical industry and in light of the fast moving biotechnology environment in which new drug candidates can emerge much faster than previously possible and become patented, the success or failure of a drug development venture is dependent on the speed and efficiency with which work can commence and proceed. One aspect of this is certainly the need for fast and accurate information. Patent information and information about stages of drug development by potential competitors are important in the initial planning stages and during the development process of a drug. From the perspective of the biotechnologist, particular areas of interest might include new methods and tools for gene expression or protein preparation, information about the structure and function relationships of proteins, and the actual protein and nucleic acid sequence information itself.

One source of such information is through the various bibliographic and substance based files available on STN International. In the following examples, we will illustrate some possible information needs and search strategies to satisfy them. A brief description of the basic protein sequence search features in the REGISTRY file is included in the appendix. Additional information can be found in the manual 'Protein Sequence Searching' published by CAS in October 1990.

II. Search Examples

EXAMPLE 1

As a simple search example consider a situation where information is needed concerning peptides having the same sequence as peptide T (ASTTTNYT or Ala-Ser-Thr-Thr-Thr-Asn-Tyr-Thr using 1- and 3-letter abbreviations, respectively) and any commercial products in which they are used. There are various strategies when conducting any search for information, but a fairly straightforward one in this instance is to begin by doing an exact sequence search in the REGISTRY file on STN as follows:

```
=> FILE REG    (enter REGISTRY File)

=> s ASTTTNYT/sqe      (search ASTTTNYT in the Exact Sequence index)

L1             8 ASTTTNYT/sqe (Answer set L1 contains 8 sequences)
```

Multiple retrievals are seen due mainly to modifications to the amino acid chain. To see the REGISTRY file record for this substance the user would ask for the record in one of several display formats, in this case one of the custom sequence display formats SQIDE3 (where the 3 indicates display in 3-letter codes):

```
=> D  ACC 106362-32-7
RN    106362-32-7
CN    L-Threonine, N-[N-[N2-[N-[N-[N-(N-L-alanyl-L-seryl)- L-threonyl]-L-threonyl]- L-threonyl]-
      L-asparaginyl]-L-tyrosyl]- (9CI) (CA INDEX NAME)
CN    Peptide T
FS    PROTEIN SEQUENCE
SQL   8
SEQ   1 Ala-Ser-Thr-Thr-Thr-Asn-Tyr-Thr
MF    C35 H55 N9 O16
SR    CA
LC    CA, CIN, CJACS, PHAR
STE   5:ALL,L
```

At this point in the search a user could continue the search to find information regarding any pharmaceutical companies developing products with this peptide. One way to do this on STN is to make use of a special feature of the REGISTRY file, namely the /LC or Registry Number Locator field. This search and display field shows pointers to other files that contain records having the specific Registry Number. In this case, the Pharmaprojects file is an excellent source for drug development information and so the search might be continued as follows:

```
=> s L1 and PHAR/LC   (search L1 answer set AND those records having
                       pointers to Pharmaprojects)

                    5714 PHAR/LC

                    L2      1      L1 and PHAR/LC
```

From this we can see that there are 5714 records in REGISTRY that contain pointers to Pharmaprojects and that our sequence query has one record in Pharmaprojects. To see this record the user then switches to the PHARmaprojects file, searches the CAS Registry Number contained in answer set L2 and then displays the record as shown:

```
=> FILE PHAR   (enter Pharmaprojects File)

=> s L2

L3             1L2

=> d L3        (display answer)

AN    6190 PHAR
CN    peptide T
CN    L-Threonine,N-[N-[N2-[N-[N-[N-(N-L-alanyl-L-seryl)-L-threonyl]-L-threonyl]-
      L-threonyl]-L-asparaginyl]-L-tyrosyl]-[CAS]
RN    ***106362-32-7***
STA   Active
CO    Originator: Non-industrial source
```

SO Pharmaprojects. PJB Publications Ltd, Richmond, Surrey, UK

TX Peptide T is an octapeptide, synthesized by Professor Lennart Wetterberg of the Karolinska Institute, Stockholm, Sweden and Dr Candace Pert at the NIH, Bethesda, Maryland, the US. It is in Phase II trials in Sweden. It is in Phase I trials in early AIDS patients in the US (Phase II trials were due in 1990) (Personal communication, Jun 1990). Bristol-Myers Squibb has discontinued development but will continue to fund a clinical trial at the University of California, the US. In a psoriasis patient, peptide T given bid iv as 1mg in 500ml saline for 28 days reduced upper body desquamation within 4 days of treatment. In a 2nd patient, the same regimen improved skin lesions and arthritis (Act Derma Venerol, 1989, 146(Suppl), 146, 117). In 20 ARC/AIDS patients receiving 0.6-3.2mg/kg/day for 4 or 12wk, followed by 8wk of 25mg/day intranasal peptide T, 4/8 with high PCR values had this value decreased (6th Int Cong AIDS (San Francisco), 1990, Abs ThA259, SB459, SB501 & SB505). The NIMH has licensed it to Integra, a research institute, which will consider joint-development with other sponsors (Scrip, 1989, 1434, 23).

DSTA World: Phase II Clinical Trial
Argentina: Available for Licensing
Australia: Available for Licensing
Austria: Available for Licensing
.
.
United States: Phase I Clinical Trial; Available for Licensing
Venezuela: Available for Licensing

CC	D5A	Antipsoriasis
	I3	Immunoreceptor binding inhibitor
	D3A	Vulnerary
	M1C	Antirheumatic, immunological
RDAT	Sep 1989	RNTE Licensing Opportunity Worldwide
	Sep 1989	Licensee discontinued Bristol-Myers

The other sequences in answer set L1 do not have pointers to the Pharmaprojects file. However, it is likely that they possess biological activities similar to that of peptide T and are under development as well. To obtain further information about these sequences, one can cross answer set L1 to the CA File and look for patent records that claim these sequences:

```
=> FILE CA
=> s L1
L4              33 L1   (33 references from patents and journals retrieved)
=> S L4 and P/DT        (Restrict answer set L4 to patents)
L5              9 L4 AND P/DT (9 patent records retrieved)
=> D 1-9 TI CS LO PY HIT        (Display title, corporate source, location,
                                publication year, and hit term)
```

L11 ANSWER 1 OF 9
TI Partially-fused pellet for sustained release of a peptide drug
CS Endocon, Inc.
LO USA
PY 1990
IT ***106362-32-7***, Peptide-T
(pellets contg. steroids and, for s.c. implantation)

L11 ANSWER 2 OF 9
TI Therapeutic methods using catalytic antibodies
CS IGEN Inc.
LO USA
PY 1989
IT ***106362-32-7***
(immunogenic transition state analogs prepn. from, in monoclonal antibodies prepn.)

L11 ANSWER 3 OF 9
TI Preparation of peptide compositions for treatment of psoriasis and neuropsychiatric disorders
CS United States Dept. of Commerce
LO USA
PY 1989
IT ***106362-34-9***
(formulation and evaluation of, for treatment of psoriasis, memory deficiency and mood disorder)

L11 ANSWER 4 OF 9
TI Peptides for treatment of psoriasis and neuropsychiatric disorders
CS United States Dept. of Health and Human Services
LO USA
PY 1989
IT ***106362-34-9***
(pharmaceuticals contg., for psoriasis and neuropsychiatric disorder treatment)

L11 ANSWER 5 OF 9
TI Sustained/controlled-release pharmaceuticals containing biodegradable polymeric carriers and water-insoluble peptides
CS Biopharm Developments Ltd. (BPD)
LO UK
PY 1989
IT ***106362-32-7*** Peptide T
(sustained/controlled-release pharmaceuticals contg. biodegradable polymeric carriers and)

L11 ANSWER 6 OF 9

TI Compositions, vaccines, and monoclonal antibodies for diagnosis, treatment, or prevention of human immunodeficiency virus (HIV) infection, and antibody production
CS Genetic Systems Corp.
LO USA
PY 1988
IT ***106362-32-7P***
(prepn. of and monoclonal antibodies to, AIDS diagnosis and treatment in relation to)

L11 ANSWER 7 OF 9
TI Threonine-containing small peptides which inhibit binding to T4 receptors and act as immunogens for preventing and diagnosing AIDS
CS United States Dept. of Commerce
LO USA
PY 1987
IT ***106362-32-7***
(AIDS virus brain and T-cell receptor-blocking)
IT ***106362-33-8*** ***106362-34-9*** ***106362-35-0***
(AIDS virus glycoprotein gp120 binding to T4 receptor in response to)

L11 ANSWER 8 OF 9
TI Threonine-containing synthetic peptides related to the human immunodeficiency virus (HIV) glycoprotein gp 120, compositions and kits containing them, and their use in diagnosis and in manufacturing medicaments for treating AIDS
LO USA
PY 1987
IT ***106362-32-7***
(AIDS virus brain and T-cell receptor-blocking)
IT ***106362-33-8*** ***106362-34-9*** ***106362-35-0***
(AIDS virus glycoprotein gp120 binding to T4 receptor in response to)

L11 ANSWER 9 OF 9
TI Small peptides which inhibit binding to T4 receptors on human T-lymphocytes, inhibition of AIDS virus infection by the peptides, and their use as immunogens
CS United States Dept. of Health and Human Services
LO USA
PY 1987
IT ***106362-32-7*** ***106362-33-8*** ***106362-34-9*** ***106362-35-0***
(human immunodeficiency virus infection of T-lymphocyte inhibition by)

Answers 3, 4, 7-9 contain peptides having modifications to the backbone sequence ASTTTNYT and exhibiting activities similar to those of peptide T. Four of the patents had been filed by the US government. The structure for the compound of answers 3 and 4 is shown below:

=> D ACC 106362-34-9 sqide3

RN 106362-34-9

CN L-Threoninamide, D-alanyl-L-seryl-L-threonyl-L-threonyl-L-threonyl-L-asparaginyl-L-tyrosyl- (9CI) (CA INDEX NAME)

FS PROTEIN SEQUENCE

SQL 8

NTE modified

type	------ location ------		description
terminal mod.	Thr-8	—	C-terminal amide

SEQ3 1 Ala-Ser-Thr-Thr-Thr-Asn-Tyr-Thr

MF C35 H56 N10 O15

STE 5:D,L,L,L,L,L,L,L

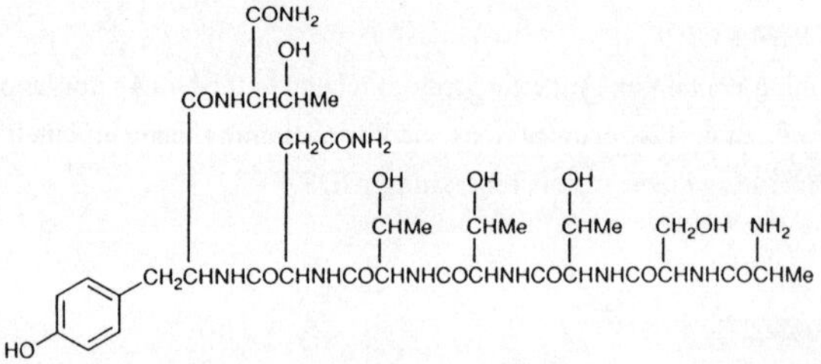

```
            CONH2
            |  OH
            |  |
        CONHCHCHMe
        |     CH2CONH2
        |     |     OH    OH    OH
        |     |     |     |     |
        |     |     CHMe  CHMe  CHMe  CH2OH  NH2
        |     |     |     |     |     |      |
HO-C6H4-CH2CHNHCOCHNHCOCHNHCOCHNHCOCHNHCOCHNHCOCHMe
```

EXAMPLE 2

This example illustrates a general search strategy one might employ in the initial stages of a drug development. A biotechnologist responsible for the design and preparation of monoclonal antibodies for cancer therapy might conduct a search in the REGISTRY file as follows:

=> FILE REG

=> E IMMUNOGLOBULIN/SSI (Expand the term "immunoglobulin" in the SSI, or substance sequence identifier.The SSI is a search field derived from the CA assigned biological name of proteins for which sequence information is available in the REGISTRY file)

E1	2	IMIDAZOLEGLYCEROL PHOSPHATE DEHYDRATASE/SSI
E2	1	IMMUNGLOBULIN G 1/SSI
E3	408 -->	IMMUNOGLOBULIN/SSI
E4	37	IMMUNOGLOBULIN A/SSI
	.	
	.	
E104	1	IMMUNOGLOBULIN Y/SSI
E105	1	IMMUNOGLOBULIN, 1-107-/SSI

E106 1 INDOLE 2,3-DIOXYGENASE/SSI

=> S e2-e5 or e9 or e11 or e12 or e59-68 or e71-76 or e78-91 or e94 or e96 or e97 or e99-e102 or e104 or e105 (Search all relevant e numbers)

L1 799 ("IMMUNGLOBULIN G 1"/SSI OR IMMUNOGLOBULIN/SSI OR . . .

There are 799 sequences in the REGISTRY file that have 'immunoglobulin' as part of their biological names. These sequences can be naturally occurring, synthetic, or genetically engineered. To examine the non-natural antibodies having potential applications for cancer therapy, one can cross the answer set L1 to the CA file and continue the search as follows:

=> FILE CA

=> S L1 AND (synthetic OR recombinant OR genetic engineer? OR genetically engineer?) AND (cancer? OR neoplasm?)

(Limit L1 to synthetic or recombinant sequences and restrict the answer set to studies involving cancer)

L2 5 L1 AND (RECOMBINANT OR GENETIC ENGINEER? OR GENETICALLY ENGINEER?) AND (CANCER? OR NEOPLASM?)(5 references retrieved)

=> D 1-5 TI DT PY (Display title, document type, and publication year)

TI In vivo tumor targeting of a recombinant single-chain antigen-binding protein
DT J
PY 1990

TI Bifunctional chimeric antibodies specific for human carcinoembryonic antigen and metal chelates
DT P
PY 1990

TI Novel recombinant, chimeric antibodies directed against a human adenocarcinoma antigen
DT P
PY 1989

TI Recombinant antibodies to Campath-1 antigen, containing foreign complementarity determining region(s), and their use in immunosuppression and cancer therapy
DT P
PY 1989

TI A genetically engineered murine/human chimeric antibody retains specificity for human tumor-associated antigen
DT J
PY 1986

To view the antibody sequences reported in the above references, one can continue the search as follows:

=> SELECT RN L2 1-5 (select for all registry numbers in references 1-5 of answer set L2)

E1 THROUGH E67 ASSIGNED

=> FILE REG (cross over to the REGISTRY file)

=> S e1-e67 AND ps/fs (Search for the registry numbers and limit the answer set to those records for which protein sequence information is available by specifying "ps", or "protein sequence", in the file segment index, /fs)

L3 29 (104491-32-9/RN OR 104491-38-5/RN OR 104491-39-6/RN OR ...
(29 sequences retrieved)

At this point, one can download all 29 sequences to a local terminal and conduct sequence similarity and other types of analyses using available software packages. Alternatively, one can continue the search on STN as follows:

=> S L3 AND 1-15/sql (Search for sequences in answer set L3 that have 15 residues or less)

L4 5 L3 AND 1-15/SQL (5 sequences retrieved)

=> FILE CA (cross the registry numbers to CA file)

=> s L4

L5 1 L4 (One reference retrieved. All 5 sequences were reported in the same reference)

=> D TI DT HIT (Display title, document type, and hit term)

TI Recombinant antibodies to Campath-1 antigen, containing foreign complementarity determining region(s), and their use in immunosuppression and cancer therapy

DT P

IT ***128096-06-0*** 128096-07-1 ***128096-08-2*** ***128096-09-3***
128096-10-6 ***128096-11-7***
(complementarity detg. region of rat YTH 34.5 HL, human recombinant antibody contg., Campath-1 antigen binding by)

It appears that we have retrieved peptide sequences that are important for the binding of antibodies to the cancer-associated Campath-1 antigen. These sequences can be displayed in the REGISTRY file:

=> FILE REG

=> D L4 1-5 SQD

SEQ LQHISRPRT

SEQ NTNNLQT

SEQ KASQNIDKYL N

SEQ EGHTAAPFDY

SEQ DFYMN

There are a number of different questions one can ask at this point. For example, would variants of these sequences have similar activities? Has anyone looked into this yet? To answer the above questions, one can conduct a subsequence family search. As an example, take the sequence DFYMN:

=> s DFYMN/sqsf not DFYMN/sqe (Search for sequences containing family equivalents of DFYMN but not DFYMN itself)

L5 971 DFYMN/SQSF NOT DFYMN/SQE (971 sequences retrieved)

To see if anyone has tested any of the of 971 sequences for cancer therapy, one can go to the CA file:

```
=> FILE CA
=> S L5 AND (cancer? OR neoplasm?)      (Search for the sequences in L5 and limit the answer
                                         set to studies involving cancer)
L6      12 L5 AND (CANCER? OR NEOPLASM?)     (12 references retrieved)
=> S L6 AND P/DT
L7      10 L6 AND P/DT       (10 out of the 12 references are patents)
```

The titles and hit terms for two of the references are displayed below:

```
TI      Antitumor agents containing activin-like polypeptides
IT      ***102524-35-6*** , Inhibin (human β.B-subunit reduced) (amino acid sequence
                   and antitumor activity of)
TI      Recombinant manufacture of cystatin C for use as virucide and neoplasm inhibitor
IT      ***116412-42-1***    ***124541-44-2***
        (amino acid sequence of and expression in Escherichia coli of synthetic gene for)
```

To view the sequences of the antitumor proteins and the regions of the sequences corresponding to the query DFYMN, one can return to the REGISTRY file:

```
=> FILE REG
=> S 102524-35-6 or 116412-42-1 or 124541-44-2      (Search for the registry
                                                     records of the antitumor proteins)
L8      3 102524-35-6 OR 116412-42-1 OR 124541-44-2
=> s L8 and DFYMN/sqsf        (Search the answer set against the query sequence)
L9      3 L8 AND DFYMN/SQSF
=> D SQIDE 1-3      (Display the sequences. Some of the display fields for the SQIDE option are
                     not shown here. Note the hit term highlighting feature which underlines the
                     family equivalents of the query sequence DFYMN)
RN      124541-44-2
CN      9-120-Proteinase inhibitor, cystatin C (human clone C6a reduced) (9CI) (CA INDEX NAME)
SQL     112
SEQ     1 LVGGPMDASV EEEGVRRALD FAVGEYNKAS NDMYHSRALQ VVRARKQIVA
        51 GVNYFLDVEL GRTTCTKTQP NLDNCPFHDQ PHLKRKAFCS FQIYAVPWQG
           =====
        101 TMTLSKSTCQ DA
HITS AT: 53-57

RN      116412-42-1
CN      Proteinase inhibitor, cystatin C (human clone pMS103 reduced)
        N-[N-(N-glycyl-L-seryl)-L-methionyl]- (9CI) (CA INDEX NAME)
SQL     123
SEQ     1 GSMSSPGKPP RLVGGPMDAS VEEEGVRRAL DFAVGEYNKA SNDMYHSRAL
        51 QVVRARKQIV AGVNYFLDVE LGRTTCTKTQ PNLDNCPFHD QPHLKRKAFC
                        =====
        101 SFQIYAVPWQ GTMTLSKSTC QDA
HITS AT: 64-68
```

RN 102524-35-6
CN Inhibin (human β.B-subunit reduced) (9CI) (CA INDEX NAME)
CN Folliculostatin (human βB-subunit reduced)
SEQ 1 GLECDGRTNL CCRQQFFIDF RLIGWNDWII APTGYYGNYC EGSCPAYLAG
=====
51 VPGSASSFHT AVVNQYRMRG LNPGTVNSCC IPTKLSTMSM LYFDDEYNIV
101 KRDVPNMIVE ECGCA
HITS AT: 15-19

It is interesting to note that the above antitumor proteins are not antibodies, yet they contain family equivalents of a complementarity determining sequence of an anticancer antibody. One can design a number of different potential anticancer drugs based on this limited information alone.

III. Beyond Protein Sequences to Nucleic Acids

Given limited resources, CAS has directed its initial development efforts to providing search and retrieval capabilities for protein sequences on STN. At the present time, most of the registered nucleic acid sequences are stored in a hardcopy file. Efforts to convert these sequences to electronic form have been initiated. The development staff are currently exploring software and hardware options for database management and searching of nucleic acid sequences.

Acknowledgements

The authors express their deep appreciation for the helpful review and comments received from Edward P. Donnell, Elizabeth L. Unruh and Patricia S. Wilson.

References

Liu-Johnson, H., Haines, R., and Hackett, W. (1990) *Biotech Forum Europe*, **8**, 204-209.

Mohindru, A, Ambrose, J. B., and Corning, J. F. (1990) Proceedings of the Montreux 1990 Conference, 219-240.

Mohindru, A., Hackett, W. F., Haines, R. C., and Corning, J. F. (1990) Online Information 90, 163-176.

Scott, J. K. and Smith, G. P. (1990) *Science*, **249**, 386-3.

Thayer, A. M. (1991) *Chemical and Engineering News*, **69**, 27-48.

Waldmann, T. A. (1991) *Science*, **2525**, 1687-1662.

APPENDIX

Protein Sequence Search Features in REGISTRY file on STN

A. Sequence search methods

There are four available options for searching sequences using amino acid codes. Each requires the corresponding field qualifier, described below. Proteins or peptides having four or more amino acid residues may be searched by a sequence query input using one-letter codes (e.g. A = Alanine), three-letter codes (e.g., 'Ala' = Alanine) or a mixture of the shorthand notations for the amino acids. There are 20 common amino acids which have both a one-letter and a three-letter code. In addition there are 59 specific uncommon amino acids that have just a three-letter code (e.g.,'Nva' = Norvaline). Use of X or 'Xxx' in a query (when permitted) will retrieve any of the 59 uncommon amino acids.

1. Exact Sequence Search (/SQE)

Exact sequence search retrieves peptides or proteins that exactly match the search query. The search query must be completely defined.

Example:

```
=> s 'Trp-Ala-Nva-Nva-Asp-Ala-Ser-Gly-Glu'/SQE          <---Search query
L1 1 'TRP-ALA-NVA-NVA-ASP-ALA-SER-GLY-GLU'/SQE          <---Answer set
=> d rn                                                 <---Display CAS
RN 85122-27-6                                               Registry Number
=> d seq3                                               <---Display sequence
SEQ3 1 Trp-Ala-Nva-Nva-Asp-Ala-Ser-Gly-Glu                  with 3-letter codes
(residues that match the query are highlighted)
```

2. Subsequence Search (/SQS)

Subsequence search retrieves exact answers plus the peptides and proteins in which the sequence is embedded. Amino acid masking and gapping (defined below) and the use of the X uncommon amino acid code are permitted.

Example:

```
=> s 'Ala-Nva-Xxx-Asp-Ala-Ser-Gly-Glu'/SQS              <--- Query
L2 1 'ALA-NVA-XXX-ASP-ALA-SER-GLY-GLU'/SQS              <--- Answer Set
SEQ3 1 Trp-Ala-Nva-Nva-Asp-Ala-Ser-Gly-Glu              <--- Sequence display
```

('Xxx' of the query sequence retrieved the uncommon amino acid Nva. The fragment corresponding to the query sequence is highlighted.)

3. Subsequence Family Search (/SQSF)

Subsequence family search retrieves exact sequences, subsequences, and answers in which family-equivalent substitution of the query amino acids occurs (see Table 1). For example, the query ADHIFC/SQSF retrieves the equivalent fragment ...PQKLYC.. .

4. Exact Family Sequence Search (/SQEF)

Exact family sequence search retrieves answers that exactly match the query and answers in which family-equivalent substitution of the query amino acids occurs.

TABLE 1: Families of amino acid equivalent

FAMILY	MEMBERS	BIOCHEMICAL DEFINITION
A (Alanine)	P, A, G, S, T	(neutral, weakly hydrophobic)
D (Aspartic Acid)	Q, N, E, D	(hydrophilic, acid amine)
H (Histidine)	H, K, R	(hydrophilic, basic)
I (Isoleucine)	L, I, V, M	(hydrophobic)
F (Phenylalanine)	F, Y, W	(hydrophobic, aromatic)
C (Cysteine)	C	(cross-link forming)

B. Amino acid masking and gapping

The following symbols may be used in sequence searches (/SQS and /SQSF) to allow for variability in the amino acid residues:

1. Amino Acid Masking Symbols (or 'wildcards')

. exactly 1 unspecified amino acid residue (which may be common or uncommon) must be present

... exactly 3 unspecified residues.

[n.] exactly "n" unspecified amino acid residues, where n is any number.

2. Amino Acid Gapping Symbols

: 0 to 1 unspecified amino acid residue per symbol.

::: 0 to 3 unspecified residues.

[n:] 0 to "n" unspecified amino acid residues, where n is any number

C. Additional capabilities

1. Combining Search Fragments

Additional search capabilities for combining the /SQE, /SQS, /SQSF and /SQEF fields are also available, using the standard Boolean operators as illustrated below.

Examples:

a. (ACWDE AND FGHIK)/SQS	Allow the two fragments to be present, in any order, and in any chain. (See the next paragraph for a definition of 'chain')
b. (ACWDE OR FGHIK)/SQS	Allow one or both fragments to be present, in any order, and in any chain.
c. (ACWDE NOT FGHIK)/SQS	Allow one fragment to not be present, in any chain.

2. Searching Multichains

If a single protein consists of two or more sequences, it is referred to as a "multichain", and each of its sequences is referred to as a "chain". One of the chains must contain at least four amino acids. Each chain may be linear or cyclic. Each chain of a multichain has a different "link value" associated with it and therefore sequence searches involving more than one fragment can be restricted to one chain using the "(L)" or link operator as illustrated in example (a) below. Similarly, the "(NOTL)" or notlink operator can be used to specify that different fragments appear in different chains as illustrated in example (b).

Examples:

a. (AWCDE (L) FGHIK)/SQS Allow the two fragments to be in any order and any number of intervening residues and require both to be in the same chain.

b. (AWCDE (NOTL) FGHIK)/SQS Allow one fragment to not be present in the same chain. Multichains have different link values in each chain.

D. Searching annotation (/NTE)

The NTE field contains sequence annotation information. This search field is divided into two types of data:

(1) *NTE Header Information* -- This includes global information about the entire sequence. Examples of header information are:

NTE multichain
cyclic, linear, cyclic
modified
homopolymer

(2) *NTE Table Information* -- This includes information that pertains to specific locations within the sequence. The NTE table is divided into three columns. The first column contains "type" information (i.e., type of data). The second contains "location" information (i.e., amino acid value, position and chain information); bridges contain both "from" and "to" locations. The third column contains "description" information, like the type of bridge. Examples of table information are:

type	------- location ------		description
terminal mod.	Ala-1	-	N-acetyl
bridge	Cys-5	- Cys-5'	covalent bridge, trimer
	Cys-5'	- Cys-5''	covalent bridge, trimer, cont.
uncommon	Aib-2	-	-
stereo	Val-1	-	D
replacement	Ser-12	-	phospha
modification	-	-	undetermined modification
modification	Phe-13	-	3-(9-acridinyl) <2; 2-6DcZ>
modification	Val-14	-	2-quinoxalinylcarbonyl <Qxc>

Example:

=> S cyclic/NTE AND dimer/NTE (Search for cyclic dimers)

L1 17 CYCLIC/NTE AND DIMER/NTE (There are 17 cyclic dimer sequences in the REGISTRY file)

E. Other search fields

In addition to the four sequence search methods described above, the user also has the option of searching several other fields as illustrated in Table 2.

In a query containing a protein sequence, only the following fields can be combined in a single search:

/SQE, /SQEF, /SQS, /SQSF, /SQL, /NTE, /FS, and /UP.

Other search fields may not be combined directly with the above fields. However, previously created answer sets can be combined with these sequence fields.

Table 2: Additional search and display fields

SEARCH FIELD	FIELD CODE	EXAMPLES
1. Sequence Related Fields		
Sequence Length	/SQL	4-20/SQL
Note (Sequence Annotation)	/NTE	MULTICHAIN/NTE
2. Name Fields		
Chemical Name	/CN	INTERFERON (HUMAN FIBROBLAST SIGNAL PEPTIDE REDUCED)/CN
Substance Sequence Identifier (or Protein Name)	/SSI	INTERFERON/SSI
Clone Designation	/CLO	PPGH-1/CLO
Gene Name	/GEN	ENV/GEN
Sulfhydryl Group	/SHG	REDUCED/SHG
Substitution	/SBN	LYSINE/SBN
Sequence Modification	/SMOD	HYDRAZIDE/SMOD
Organism Name	/ORGN	HUMAN IMMUNODEFICIENCY VIRUS/ORGN
Sequence Name Segment	/SQNS	IMMUNODEFICIENCY VIRUS/SQN
Sequence Name Segment	/SQNS	VIRUS/SQNS
3. Miscellaneous Fields		
Basic Index	/BI	PEPTIDE/BI
Chemical Class Identifier	/CI	MAN/CI
File Segment	/FS	PROTEIN/FS
Molform	/MF	C61H78Br2N8O14/MF
Sequence Field Availability	/SFA	SEQ/SFA
Source of Registration	/SR	CA/SR
Update Date	/UP	UP>900933

International collaboration of protein databases: sequence databases, nonsequence databases and variant databases

Akira Tsugita and Fuminori Okibayashi

JIPID, Research Institute for Biosciences, Science University of Tokyo, Yamazaki Noda 278, Japan

Several lines of developing methodologies have been resulting in data accumulation in biological field. Especially, biotechnology including protein engineering and human (and the other) genome projects have become popular. International collaboration among those collecting organisation and data dissemination is indispensable. The International advice committee for nucleic acid sequence databanks, the PIR-International association for protein sequence databases, and human genome organisation, HUGO are acting and catalysing international collaboration in the respective fields.

CODATA, the committee on data for science and technology of the International Council of Scientific Unions (ICSU) set up Task Groups of Hybridoma Data Bank and the Microbiological Strain Data New Work, 1982 and 1984 respectively. In 1984, yet another important Task Group on Protein Sequence Data Banks which has been expanded and changed its name to the Task Group on Biological Macromolecules (BMM) since 1988. Two other international task groups were started in biology field at 1988, Systematic Nomenclature for Foods in Numerical Data Banks and Standardised Terminology for Access to Biological Data Banks (STAB). Among the above, two activities are summarised below [1].

CODATA Task Group on biological macromolecules (BMM)

The Task Group (chaired by B. Keil (1988–89) and then by A. Tsugita (1989-) is invaluable in coordinating existing databanks and encouraging emerging databanks. The Task Group covers Protein Sequence Data Banks (PIR- International; NBRF, MIPS, JIPID), Nucleic Acid Sequence

Data Banks (GenBank, EMBL, DDBJ), the X-ray crystallographic Data Bank (PDB), the NMR DataBank, the Carbohydrate Database (Carb Bank), Hybridoma Data Bank (HDB) and Standardised Terminology Activity (STAB). Among them two Protein Sequence Data Banks, MIPS, JIPID, protein NMR DataBank [2] and the European node of Carb Bank were initiated by encouragements by the Task Group. The Task Group established a standard format for exchanging sequence databases [2]. Moreover, the Task Group endeavored to interlink the related databases such as sequence-structure database [6,7] and encourage creation of protein biological activity and physicochemical property databases [8], variant database [9], as well as Protein 2D database.

CODATA Commission on standardised terminology for access to biological data banks (STAB)

The commission has aimed to improve international access to and use of information resources in biology promoting international and interdisciplinary cooperation in the use of terminology controlled vocabularies. The development and use of terminology controlled vocabularies is a task which involves: a) The identification of existing committees and *ad hoc* groups working on vocabulary and nomenclature. b) The publication and wide international dissemination of information of these committees. c) The identification of gaps in standardised nomenclature and terminology in specific subdisciplines of biology and the encouragement to provide them in these area. d) Development of systems for the storage and retrieval of comprehensive data on the source, definition, and hierarchical relationships of biological terms, etc. e) The study of the principles of standards for terminology and their application to an integrated system for biological vocabulary. The Commission has Lois Blaine as Chair and J. Franklin and A. Tsugita as Vice-Chairs.

Protein sequence database

By encouragement of the Task Group BMM, a protein data bank called Japanese International Protein Information Database (JIPID) was founded at Noda campus, Science University of Tokyo in 1987. Another protein data bank, Martinsried International Protein Database (MIPS) was also founded at Max-Planck Institute for Biochemistry, by Munchen, Germany in 1988. Together with National Biomedical Research Foundation (NBRF) at Georgetown University Medical Center, Washington, D.C., these two data bases have formed a new trifoliate association which has been producing an unique protein sequence data base, PIR-International. Each nodes collect and disseminate the protein sequences, geometrically covering and create a single data file on the basis of the CODATA standardised format [3].

JIPID is Asian and Oceania node, while NBRF covers Northern and Southern America and MIPS covers Europe. JIPID has collaborative several national subnodes; Australia (Ludwig Institute for Cancer Research, Melbourne), China (Institute of Biophysics Chinese Academy of Science, Beijing), Taiwan (Institute of Zoology, Academia Sinica, Taipei), India (Bioinformatics University of Poona, Pune), Pakistan (HEJ Research Institute of Chemistry University of Karachi, Karachi), and Korea (Korean Advanced Institute of Science and Technology, Seoul). The trifoliate protein nodes closely collaborate with the corresponding trifoliate nucleic acid databases, DDBJ (Japan), EMBL (Germany) and GenBank (USA). These six databases, nucleic acid and protein sequences, set up an identical data direct submission form and the submitted data are shared among these databases. Each protein node solely or collaboratively has developed specialised activities and the resulting data are shared among the three protein nodes. For example, MIPS has initiated Yeast genome activity by the support of EEC project [4,5], nonannotated sequence data (INQ) and a key words databank and the other two nodes are joining these activities. NBRF collaborates with protein tertiary structure databank (PDB) to form a linking data base between sequence data and structure data, called NRL_3D [6]. NBRF developed a multi database program XQS, which can operate maximum 16 protein and nucleic acid datafiles. JIPID developed two dimensional data base (PSS) [7] using PIR and PDB. And JIPID has been developing biological activity data base (BAD), physicochemical property data base (PPD) [8] and variant data base (VAD) [9]. JIPID also made E. coli genome data base [10] with NBRF and T4 phage genome data base. Rice genome data base is a collaborate project between JIPID and NBRF. JIPID is as well the Japanese node of carbohydrate databank (CarbBank).

Recently the design of database has changed to improve providing a complete and up-to-date data set of protein sequences. These changes have resulted in a restructuring of the sequence database sections in PIR. These sections are 'Annotated Entries (Main)', 'Preliminary Entries (New), which are not finally edited' and an additional section 'Unverified Entries (INQ).' The last new section INQ is a minimally necessary to make the sequence data useful; and corresponds to a single sequence as published in a single manuscript. Each entry contains only an entry identification code, an entry title, an optional species line, the reference citation with the single reference number, an optional cross reference line, and the sequence. This restructure increased the number of data files from 16,500 sequences to 28,000 sequences in six months. Updates of PIR international at March 1991 is a total of 28,232 sequences, 8,076,497 residues, including 7,967 Annotated Entries, 12,607 Preliminary Entries and 7,568 Unverified Entries.

Database web

The protein sequence data, Biological Active Database, Physicochemical Database and Variant Database are made in a similar format and share quite a few data items in common. The data items in each independent databases are noted in boxed type and the normal letters shows common items in the databases in Table 1. We made these data structures in a database web to transfer from one to the other, avoiding redundancy. Protein Secondary Structure Database is also included in the same database web. The multidatabase information retrieval program called XQS are now available. This program merges the capabilities of PSQ, NAQ and the above mentioned databases and is designed around a single index structure that dramatically decreases the information retrieval time of the program while providing simultaneous access to multiple databases. Included in the release is software that allows indexes from the GenBank(R), EMBL, or other databases to be merged into the index. Fig.1 describes the schematic relation of the database web.

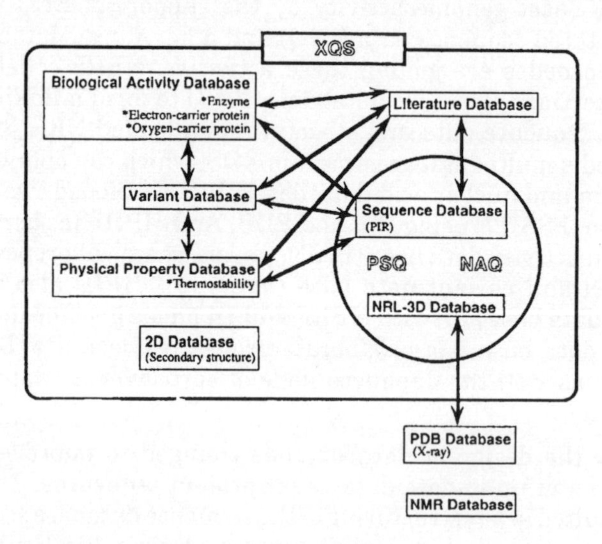

Figure 1: Database Web

Biological Activity Database (BAD) [8]

Protein has a variety of biological activities. In 1988 JIPID initiated to build the biological activity database BAD to give proper annotations of biological activities and it allows to make comparisons between those of the wild protein and variant proteins in variant data file. The database is divided into two parts, General and Specific databases. The General database covers the common items, such as SYSTEMATIC NAME, REACTION (mechanism) and FUNCTION for a group of proteins, and REFERENCES to the specific individual data files. Specific data file covers the biological activities of 'biologically active proteins', sequences of which has already been stored in PIR. It should be noted that the unit of data file is the matured biologically active protein but not for the premature chain, like in The sequence data. The ORGANISATION data

PIR(PSQ)	enzyme	physicochemical	variant
Header	Header	Header	Header
Entry-code	Accession number	Accession number	Accession number
Title	Title	Title	Title -V *
EC-number	EC-number	EC-number	EC-number
	Systematic-name		
Alternate-name	Alternate-name	Alternate-name	Alternate-name
Includes			
Reference	Reference	Reference	Reference
Source	Source	Source	Source
Host	Host	Host	Host
			Method
			Expression
Comment			
	Organisation	Organisation	Organisation-V
	Reaction	Reaction	Reaction
	Function		Function
		Physicochemical	Physicochemical
	Post-translational		
			Pathology
	Application		Application
Genetic	Genetic		Genetic
Keywords			
Feature	Feature	Feature	Feature
			Variation
Summary	Summary	Summary	Summary-V **
Sequence	Sequence	Sequence	Sequence-V **
Cross-reference	Cross-reference	Cross-reference	Cross-reference

Normal type: transferred from the other datafiles with or without minor modifiecations

Boxed type: originl input data

* transfer with manual modification

** transfer with modification by program

Table 1 Database Web

item provides the structural organisation of the mature protein. It consists of the number of each subunit, the name of subunit, and residue numbers of the mature polypeptide followed by the PIR entry code. By this item, the mature sequence of the composed subunits may be automatically displayed and residue numbers in Sequence and Feature table are altered. FUNCTION is also an important data item which denotes the functions of the protein. The other data items, TITLE, SOURCE, HOST, REACTION, FEATURE, GENETIC, SEQUENCE, etc., are taken, with or without minor modifications, from the PIR sequence or the BAD General databases. At the initial stage, we have focused on Enzyme, Electron Carrier Protein, and Oxygen carrier protein in biological function. The Enzyme data file will be publically released in 1991. Examples of general data file and specific data file list in Table 2 and 3 respectively.

Physicochemical Property Database (PPD)

The Physicochemical Property Database is the second non-sequence database and contains information on physicochemical properties of proteins. Initially the entries in this database are confined to proteins that have entries in the protein sequence database. The data currently entered concern with thermostability, pI, and optical data.

The TITLE and HOST, SOURCE, ORGANISATION, COFACTORS, FEATURE, REACTION, and OPTIMAL-PH data items are taken from the PIR or the BAD. Thermostability data items include TRANSITION-NUMBER, which states the number of transition states. For each transition the TRANSITION TEMPERATURE, ENTHALPY change, HEAT CAPACITY and FREE ENERGY change are given. These items are followed by the conditions under which the transition state was measured experimentally. As in the Biological Activity Database, the SEQUENCE data item refers to the sequence of the biologically active mature protein. Other physicochemical data are described by the data items ISOELECTRIC-POINT and SPECTRUM for the pI and spectral data, respectively. The format of PPD is shown in Table 4.

The Variant Database (VAD) [9]

Recent advances in gene manipulation has resulted in an increasing number of reports concerning the generation and identification of variant biomacromolecules, artificially created or naturally occurring variants. There is much interest, both academic and application, in using biological and physicochemical data of such variants to elucidate structure-function relationships, and to develop novel compounds with specified properties. To fully utilise these growing data, it is necessary to be able to compare variant molecules with each other as well as with wild type molecules. The Variant Database (VAD) aims to serve as a repository for data concerning both variations of polypeptide sequences and biological activities and physicochemical properties of variant polypeptides and

ACCESSION-NUMBER	**KG0039**
TITLE	Histidine decarboxylase
EC-NUMBER	4.1.1.22
#Enzyme-Group	Lyase
#Subclass	Carbon-carbon lyase
#Description	Carboxyl-lyase
SYSTEMATIC-NAME	L-Histidine carboxy-lyase
DATE	1-OCT-88
#Revised	12-SEP-90
LITERATURE	
AUTHORS	Boaker, E.A., Snell, E.E.
JOURNAL	The Enzymes (3rd ed.) 6, 217-253, 1972
LITERATURE	
AUTHORS	Aures, D., Hakanson, R.
JOURNAL	Meth. Enzymol. 17B, 667-677, 1971
DISTRIBUTION	Bactera, plants and animals
ORGANIZATION-G	A single chain (eukaryote) Oligomer of two different chains (bacteria)
COFACTORS-G	Pyridoxal 5'-phosphate (eukaryote enzyme) Pyruvoyl residue covalently linked to N terminal (bacterial enzyme)
REACTION	L-Histidine = histamine + CO2
ASSAY	Manometric determination of CO2 or fluorometric determination of histamine.
FUNCTION	
#Substrate-G	Highly specific to L-histidine
#Inhibitor-G	Some imidazole compounds as imidazole propionic acid and urocanic acid
#Optimal-pH	pH 4 to 5
COMMENT-G	Eukaryote enzyme has pyridoxal phosphate while prokaryote enzyme has pyruvoyl N-terminal as the cofactor, respectively. Structural similarities are scarcely observed between the two families.
CROSS-REFERENCES(S)	KS1009 Lactobacillus sp. <DCLBHP> KS1010 Lactobacillus buchneri <DCLBHB> KS1025 Clostridium perfringens <DCCLHP>

///

Table 2 Display of enzyme database (general)

ACCESSION-NUMBER	**KS1009**
TITLE	Histidine decarboxylase (EC 4.1.1.22) - Lactobacillus sp.
SYSTEMATIC-NAME	L-histidine carboxy-lyase
DATE	17-MAY-91
SOURCE	Lactobacillus sp. (Strain 30a, ATCC 33222)
LITERATURE	KS1009
AUTHORS	Rosenthaler, J., Guirard, B.M., Chang, G.W., and Snell, E.E.
JOURNAL	Proc. Nat. Acad. Sci. USA 54, 152-, 1965
TITLE	
ORGANIZATION	6 small; 1-81 [DCLBHP], 6 large; 83-307 [DCLBHP]
REACTION	L-Histidine = histamine + CO2
FUNCTION	
Substrate	Highly specific to L-histidine
Inhibitor	Some imidazol compounds
Optimum-pH	4.8
Activity	
#Substrate	L-Histidine
#pH	4.8
#Temparature	37 C
#Km	0.9 mM
#Comment	Km value depends on the pH and the ionic strength.
#Substrate	L-Histidine
#pH	8.4
#Temperature	37 C

Competitive inhibitor	Ki(mM)
Imidazole propionate	1.8
Urocanate	2.1
Imidazole	3.2
N-Methylimidazole	7.2
Pyridine	1.4 <ROS>

POST-TRANSLATIONAL	The proenzyme is a hexamer of identical chains (pi chains). Each pi chain cleaves nonhydrolytically to form the small chain and the large chain. A serine residue preceding the large chain is converted to a pyruvoyl group covalently bound to 1-Phe of the large chain.
FEATURE	
large chain	
Residues	Feature
1	Modified site: blocked amino end (Phe), by pyruvoyl group

continued over page

SEQUENCE-P

DCLBPH

small chain

Mol. wt. unmod. chain = 8,841 Number of residues = 81

```
                5             10            15            20            25            30
1    S E L D A K L N K L G V D R I A I S P Y K Q W T R G Y M E P
31   G N I G N G Y V T G L K V D A G V R D K S D D D V L D G I V
61   S Y D R A E T K N A Y I G Q I N M T T A S
```

DCLBPH

large chain

Mol. wt. unmod. chain = 24,820 Number of residues = 225

```
                5             10            15            20            25            30
1    F T G V Q G R V I G Y D I L R S P E V D K A K P L F T E T Q
31   W D G S E L P I Y D A K P L Q D A L V E Y F G T E Q D R R H
61   Y P A P G S F I V C A N K G V T A E R P K N D A D M K P G Q
91   G Y G V W S A I A I S F A K D P T K D S S M F V E D A G V W
121  E T P D E D E L L E Y L E G R R K A M A K S I A E C G Q D A
151  H A S F E S S W I G F A Y T M M E P G Q I G N A I T V A P Y
181  V S I P I D S L P G G S I L T P D K D M E N L T M P E W L E
211  K M G Y K S L S A N N A L K Y
```

Table 3 Display of enzyme database (specific)

ACCESSION NUMBER	
TITLE	
EC-NUMBER	
ALTERNATIVE-NAME	
DATE	
SOURCE	
HOST	
ORGANIZATION	
COFACTORS	
FEATURE	
REACTION	
OPTIMAL-PH	
TRANSITION-NUMBER	Number of transition states. Value and/or description
TRANSITION-STATE-N	Transition state
#Trn-STEP-N	Transition temperature in K (C) or transition concentration in molar (M)
#DELTA-G-25-STEP-N	Gibbs energy change of transition at 25 C in J/mol (kcal/mol)
#DELTA-H-25-STEP-N	Enthalpy change of transition at 25 C in J/mol (kcal/mol)
#DELTA-S-25-STEP-N	Entropy change of transition at 25 C in J/mol/deg (kcal/mol/deg)
#DELTA-CP-25-STEP-N	Heat capacity change of transition at 25 C in J/mol/deg (cal/deg/g)
CONDITIONS	–
#Methods	–
##DSC(a)	Differential scanning calorimetry, adiabatic
##DSC(na)	Differential scanning calorimetry, non-adiabatic
##vH	van't Hoff equation
##FC	Flow calorimetry
##ITT	Isothermal titration
#Denaturant	Denaturant and concentration used
#Temperature	Temperature in K (C) used in the experiment
#Reversibility	Description such as reversible, irreversible, or not described
#pH	Value
#Ionic-Strength	Value in M
#Buffer	Description plus molarity
#Salt	Description
#Sample-Concentration	Value or range in mM (mg/ml)
COMMENT – PHYSICAL	
SPECTRUM	
#Wavelength	Value in nm
#Extinction-Coefficient	Value and/or description
2D-GEL	
#app-Mol. wt.	
#ISOELECTRIC-POINT	Value
SUMMARY	
SEQUENCE-P	
CROSS-REFERENCE	
///	

Single type; transferred from the other datafiles with or without minor modifications
Bold type; originally input data

Table 4. Proposed format for the physico-chemical properties database

ACCESSION M0009
TITLE Lysozyme (EC 3.2.1.17) variants - Human
REFERENCE M0009001
AUTHORS Muraki, M., Morikawa, M., Jigami, Y., and Tanaka, H.
JOURNAL Biochim. Biophys. Acta 916, 66-75, 1987
TITLE The roles of conserved aromatic amino-acid residues in the active site of human lysozyme: a site-specific mutagenesis study.
SOURCE Homo sapiens (man)
METHOD
DNA: synthetic gene, cassette mutagenesis [M0009001][M0009002][M0009003]
DNA: synthetic gene, gapped duplex mutagenesis [M0009004][M0009005][M0009006]
EXPRESSION
Vector M13mp11
Host Saccharomyces cerevisiae strain KK4
ORGANIZATION 1 ; 1-130 <P1 ; LZHU>
REACTION Hydrolosis of 1,4 beta-linkages between N-acetylmuramic acid and N-acetyl-D-glucosamine residues in a peptidoglycan and between N-acetyl-D-glucosamine residues in chitodextrin
FUNCTION
#Kinetic parameter for bacterial substrate
#Substrate: Micrococcus lysodeikticus
#pH 6.2, phosphate buffer, 25 C

	Km,app (ug/ml)	**Vmax (deltaA/min/ug)**	**Vmax/Km,app**
wild-type	67	0.28	4.2E-3
M0009001	–	–	–
M0009002	120	0.34	2.8E-3
M0009003	190	0.051	2.7E-4
M0009004	120	0.32	2.7E-3
M0009005	74	0.068	9.2E-4
M0009006	55	0.069	1.3E-3

FEATURE
6-128,30-116 Disulfide bonds:
35 Active site: Glu
53 Active site: Asp
VARIATION
M0009001 1-34,'D',36-130
M0009002 1-62,'L',64-130
M0009003 1-63,'Y',65-130
M0009004 1-63,'F',65-130
M0009005 1-108,'Y',110-130
M0009006 1-108,'F',110-130
SUMMARY
Mol. wt. unmod. chain = 14,701 Number of residues = 130

Composition

14 Ala A	6 Gln Q	8 Leu L	6 Ser S
14 Arg R	3 Glu E	5 Lys K	5 Thr T
10 Asn N	11 Gly G	2 Met M	5 Trp W
8 Asp D	1 His H	2 Phe F	6 Tyr Y
8 Cys C	5 Ile I	2 Pro P	9 Val V

SEQUENCE

```
                5              10             15             20             25             30
1    K  V  F  E  R  C  E  L  A  R  T  L  K  R  L  G  M  D  G  Y  R  G  I  S  L  A  N  W  M  C
31   L  A  K  W  E  S  G  Y  N  T  R  A  T  N  Y  N  A  G  D  R  S  T  D  Y  G  I  F  Q  I  N
61   S  R  Y  W  C  N  D  G  K  T  P  G  A  V  N  A  C  H  L  S  C  S  A  L  L  Q  D  N  I  A
91   D  A  V  A  C  A  K  R  V  V  R  D  P  Q  G  I  R  A  W  V  A  W  R  N  R  C  Q  N  R  D
121  V  R  Q  Y  V  Q  G  C  G  V
```

Table 5. Display of variant database (multiple display)

polynucleotides. The nature of the database requires all data obtained from the current literature or received from the primary investigator. There are also many variants applied for patents. The unit of VAD is an individual altered protein.

VAD has TITLE data item composed of the name of wild type molecule donation of sequence modification and biological source of the molecule. Among the data records are several novel data items; METHOD (to produce mutant), EXPRESSION (denoting vector and host to express the variant gene), ORGANISATION (as the BAD with substitution, insertion, deletion and fragment notation with the residues), FUNCTION (comparable with the wild type values in the BAD), PHYSICOCHEMICAL (comparable with the wild type values in the PPD), PATHOLOGY (having several specified sub-items), APPLICATION, etc. One of the VAD data file is shown in Table 5.

Protein Secondary Structure Database (PSS)

A new protein secondary structure database (PSS) was developed to correlate the protein sequence database of the PIR with the atomic coordinates and bond connectivities database of the Protein Data Bank (PDB) in Brookhaven National Laboratory. The database includes secondary structures determined by X-ray diffraction analysis, but not predicted secondary structures. Users may select the secondary structure information from either PDB or the output of the program of 'Define Secondary Structure of Proteins' made by Kabsch and Sander. The present database adopts the PIR format and XQS programs. Secondary structure(s) of structural interest are displayed together with protein primary structures. Secondary structure of a desired length of peptide fragment is displayed upon request, as are the peptide fragments that correspond to a defined secondary structure. The PSS database is now publically available and example of the database is shown in Table 6.

Sequence-structure database (NRL_3D)

A new sequence-structure database, NRL_3D, has been developed from sequence information extracted from the Brookhaven Protein Data Bank (PDB) and these data have been correlated with those of the protein sequence database of the PIR.

There are distinct differences between the representation of sequence data in the PDB and PIR. In the PDB, multiple chains are often stored in single entries and the sequence numbering system employed often contains negative numbers and insertion codes. As the majority of sequence search, analysis and retrieval programs are designed to work on data formatted similarly to that in the PIR, the sequence data in the PDB are not readily accessible to existent sequence analysis software. Perhaps more disturbing is that, as has been reported by Lesk et al., there are often disagreements between corresponding sequences in the two databases. As the crystal structure refinements are often based on

LS0119
Alcohol dehydrogenase (EC 1.1.1.1) E chain (NAD, DMSO, Zn++)
- Horse

Cross-reference: #PDB: 6ADH

Organization: 2 E;1-374<DEHOAL>

Eklund, H., Samama, J.-P., Wallen, L., Branden, C.-I.,Akeson, A., and Jones, T.A., J. Mol. Biol. 146, 561, 1981
Title: Structure of triclinic ternary complex of horse liver alcohol dehydrogenase at 2.9 angstroms resolution.

Cofactor: NAD; dimethyl sulfoxide (DMSO); Zn++

Residues	Feature
46,48,67,93, 141,174	Binding site: DMSO, E chain <COF>
47,51,178,198-200, 202-203,222-224, 228,269-271, 292-294,319, 369	Binding site: NAD, E chain <COF>
46-53,168-175, 176-188,201-215, 229-236,250-258, 271-282,305-311, 324-338,355-365	Helix: H1,H2,HA,HB,HC,HCD,HE,HS,H3,H4, E chain <hlx>
156-160,88-92, 68-71,41-45, 369-374,347-352	Sheet S1: E(-)E(-)E(-)E(-)E(+)E, E chain <brr>
86-87,72-78, 34-40,148-152	Sheet S2: E(-)E(-)E(-)E, E chain <sht>
9-14,22-29, 129-132,135-138, 62-64	Barrel S3: E(-)E(-)E(-)E(+)E(+), E chain <sht>
238-240,218-224, 193-200,263-269, 287-293,312-318	Sheet S4: E(+)E(+)E(+)E(+)E(+)E, E chain <sht>
32-35,80-83, 84-87,118-121, 132-135,140-143, 153-156,165-168, 190-193,244-247, 258-261,283-286, 295-298,319-322, 338-341,344-347, 351-354	Turn:T1,T2,T3,T4,T5,T6,T7,T8,T9,T10,T11, T12,T13,T14,T15,T16,T17, E chain <tur>
1-8,15-21,30-31, 54-61,65-67, 79,93-117,122-128, 139,144-147, 161-164,189, 216-217,225-228, 237,241-243, 248-249,262, 270,294,299-304, 323,342-343, 366-368	Coil: E chain <coi>

continued over page

```
    5    10   15   20   25   30
1 S T A G K V I K C K A A V L W E E K K P F S I E E V E V A P
              >>>>S31b>>>             <<<<<<S32b<<<<<

31 P K A H E V R I K M V A T G I C R S D D H V V S G T L V T P
      >>>>>S23>>>>> <<<S14<<< <- - - - H1- - - ->
 <- T1- - >

61 L P V I A G H E A A G I V E S I G E G V T T V R P G D K V I
 <S35b<        >>S13>> <<<<<S22<<<<< <-T2-> >S21><<<S1
                                        <-T3 -->

91 P L F T P Q C G K C R V C K H P E G N F C L K N D L S M P R
 2<<<                                                  <- T4-

121 G T M Q D G T S R F T C R G K P I H H F L G T S T F S Q Y T
 ->           >>S33b>    <<S34b<  <- T6 ---->      <<<S2
                   <- T5 - ->

151 V V D E I S V A K I D A A S P L E K V C L I G C G F S T G Y
 4<<<      >>>S11>>>               <------ H2 -----> <---------
     <-T7->                <-T8-->

181 G S A V K V A K V T Q G S T C A V F G L G G V G L S V I M G
 --HA----------------->      >>>>>>S43>>>>>> <----------- HB---------
                   <- T9-->

211 C K A A G A A R I I G V D I N K D K F A K A K E V G A T E C
 --------------->    >>>>>S42>>>>>>        <------- HC------>  >S41>

241 V N P Q D Y K K P I Q E V L T E M S N G G V D F S F E V I G
       <-T10->     <------- HCD------>          >>>>>S44>>>>
                                  <-T11->

271 R L D T M V T A L S C C Q E A Y G V S V I V G V P P D S Q N
  <--------- HE-------------->  <-T12-> >>>>>>S45 >>>>>>    <-T13->

301 L S M N P M L L L S G R T W K G A I F G G F K S K D S V P K
          <------ HS-----> >>>>> S46 >>>>>> <-T14->      <----------------

331 L V A D F M A K K F A L D P L I T H V L P F E K I N E G F D
 H3------------------->                >>>>S16>>>>      <--------- H4
               <-T15->      <-T16->      <-T17->

361 L L R S G E S I R T I L T F
 -------------->      >>>> S15 >>>>
```

Table 6. Display of protein secondary structure database

prior knowledge of the sequence, these discrepancies are not easily resolved.

Sequence data from selected coordinate sets (those with resolutions of 2.5 A or better and corresponding to well-defined sequences) in the PDB have been extracted and restructured; the PIR or XQS software has been extended to access this information and to handle transparently the numbering system conversions. This software interfaces with standard molecular modeling programs and allows the three dimensional structure of identified sequences to be displayed directly. These data have been directly cross-referenced to the PIR.

This work will form the basis of a direct cross-linkage between the PIR and the PDB. In collaboration with the Brookhaven National Laboratory, crosslinks to the PDB will be directly incorporated into the PIR in a form that will allow the exact regeneration by computer of the sequence as reported in the PDB with its original numbering system. In collaboration with the staff at the new NMR Structure Database, similar linkages will also be established with this database. The database are distributed in the PIR-International tape.

References

1). Tsugita, A., Okibayashi, F., Kunisawa, T., and Satake, K. Trends of biological database in CODATA and protein information database in Japan, *CODATA Bull* (in press)

2). Ulrich, E.L., Markley, J.L., and Kyogoku, Y. *Protein Seq. Data Anal.* **2**, 23-37, 1989

3). George, D.G., Mewes, H.W., and Kihara, H. A standardise format for sequence data exchange, *Protein Seq Data Anal* **1**, 27-39, 1987

4). Tsugita, A., and Barker, W.C. Information of Task Group of Biological Macromolecule, *CODATA Bull* (in press)

5). Sgouros, J.G. The European Yeast Genome Sequencing Network; The complete sequence of chromo some III from the yeast Saccharomyces, Cold Spring Harbor Meeting on Genome Mapping and Sequencing. Cold Spring Harbor Lab. New York pp. 248 (1991)

6). Pattabiraman, N., Namboodiri, K., Lowrey, B.P., and Gaber, B.P. NRL_3D: a sequence-structure database derived from the protein data bank (PDB) and searchable within the PIR environment. *Protein Seq. Data Anal.* **3**, 387-405, 1990

7). Suzuki, H., Kolasker, A.S., Samuel, S.L., Otsuka, J., and Tsugita, A. A protein secondary structure (PSS) data base. *Protein Seq. Data Anal.* (in press)

8). Jone, C.S., Tsugita, A., Satake, K., Okibayashi, F. Non-sequence database for biological activity and physicochemical properties. *Protein sequence Data Anal.* (in press)

9). Ubasawa, A., Okibayashi, F., Jone, C.S., Ikehara, M., George, D.G., and Tsugita, A. A variant database. *Protein Seq. Data Anal.* (in press)

10). Kunisawa, T., Nakamura, M., Watanabe, H., Otsuka, J., Tsugita, A., Yeh, L.S., George, D.G., and Barker, W.C. Escherichia coli K12 genomic database. *Protein Seq. Data Anal.*, **3**, 157-162 (1990)

11). Lesk, A.M., Boswell, D.R., Lesk., V.I., Lesk, V.E., and Bairoch, A. A cross reference table between Protein Data Bank of macromolecular structures and the National biomedical Research Foundation-Protein Identification Resource amino acid sequence data bank. *Protein Seq. Data Anal.* **2**, 295-308, 1989

12). Tsugita, A., Ubasawa, A., Jone, C.S., Ikehara, M., Chen, H.R., and George, D.G. An artificial variant database and data web. 'Protein Enginering: Protein Design in Basic Research, Medicine and industry' (ed by Ikehara, M.) Springer-Verlag, pp. 273-278

USDA, Plant Genome Research Program: problems and solutions

J.P. Miksche

U.S. Dept. of Agriculture, BARC West, Bldg. 005, Beltsville, MD 20705, USA

M.K. Berlyn

Dept. of Biology and School of Forestry, and Environmental Studies, 355 Osborn Laboratory, Yale University, New Haven, CT 06520, USA

S.R. Heller

U.S. Dept. of Agriculture, BARC West, Bldg. 005, Beltsville, MD 20705, USA

Biotechnology products are not in the market place as promised by the hype of the middle and late 1970s. Gene mapping research will help in meeting those promises, because the genes must be found first before they can be used. As a result of gene mapping and associated genomic research, voluminous information will be generated and it must be recorded, stored and made readily accessible to others. In anticipation of the increased information, planning prototype analysis, and experimentation to establish a relational data base for all plant species as related to the USDA Plant Genome Research Program is underway.

Plant science applied to agriculture harnesses plant's potential by helping to insure an adequate supply of agriculturally based food, fiber, industrial products, and clean air for the earths ever-growing population. Genetics and other sciences have played a major role in agriculture over the years. Humans have improved cultivated plants by selection over several millennia and more scientifically for the past one hundred through application of the principles of heredity. The use of these genetic principles enabled the rapid improvement of many crop varieties in the twentieth century through hybridization and selection techniques. Early progress produced substantial increases in yield, pest resistance tolerance to environmental stresses, and other genetic traits of economic importance. But the giant steps are getting rarer, and traditional plant breeding technologies in some crops are reaching limits. Efficient methods to identify, isolate, and transfer desirable genes need development to overcome these limitations of conventional breeding methods.

Since the late 1970s, the rapidly evolving tools of molecular biology have promised further improvement in crop and forest species. A decade of genetic engineering research, however has yet to yield the full potential of molecular biology as related to crop improvement.

The major problem facing agriculture today in the developed countries is production efficiency. A strong need to reduce grower input cost with concomitant increase in yield and quality with reduced environmental impact is mandatory. Solution of this agricultural challenge hinges upon the tools of the 'new biotechnology'. Recognition of this is manifested by the world wide interest and thrust of the public and private sectors. The major countries of the world have several large and small firms investing resources in the field, an increase significantly higher than observed in 1987. Despite major 'buy-outs,' shifts, and business failures, there has been an increase in the number of firms entering the competition since 1980. It is evident, therefore that interest in advancements of biotechnology and associated products for the market place is not declining. However, the release of new plant varieties 'for sale' via biotechnology applications has not occurred.

The key points of this paper are, 1) A need to locate genes that code for desirable characteristics to produce plants that possess disease, heat, cold, drought tolerances and other yield deterrents is pressing. These characteristics determined by genomic mapping research will increase farming efficiency in yielding food and non-food products for the consumer. 2) Genomic research, associated molecular biology techniques and other related to agricultural sciences will generate much information that requires input, storage, retrieval and exchange by users.

Problems and solutions

The problems before us in agriculture are complex scientifically, economically, politically, and they are intimately linked with sociological changes through altering agricultural practices and policies. It is not possible to specifically cover all of the above problems, but a general panorama of the problems portray familiar scenes and challenges of increased world population with related food/fiber needs, which in turn, is directly related to efficient crop productivity, quality, increased petroleum energy cost, related petrochemical production fertilizer, environmental quality and improving the economic base of agricultural products. Specifically, the major problem facing agriculture today is plant production efficiency.

A strong need to reduce grower input cost with concomitant increase in yield and quality with reduced environmental impact is the need. This need will take a rather long time to reach. Considerable research in plant science must be done to overcome the impediments of biotechnology product advancement and development.

Today, the precise tools afforded by molecular biology and genetic engineering can aid in determination of factors or key molecules that regulate carbohydrates, proteins, and fats. Opportunities in this field include changing chemical composition of the plant product, improving processing quality, enhancing resistance to stress altering plant size, and changing the ratio of the nutrient distribution in grain, fruit, leaves, stems or roots. However, many of the genes related to growth form and yield are multiple and operate in a complex developmentally regulated fashion. The significance of a complex integrated gene system is difficulty in gene transfer compared with single gene constructs and delivery. The related enzyme systems and metabolic pathways coupled with complex gene cascade systems presents a serious scientific problem to unravel.

Elucidation of the biochemistry and metabolism of several key compounds and pathways in higher plants is mandatory to implement effective transfer of complex genes. This is particularly relevant to directing the regulatory enzymes systems that are related to partitioning and distribution of desirable carbohydrates, lipid, and protein molecules to specific plant organs. The emphasis of many studies, for example, is directed toward the movement of lipids and proteins to filling soybean seeds, sugar molecules to sugar beet root, and rubber to leaves or stems of guayule or hevea.

Indispensable to gene transfer of desirable single and complex gene systems are plant cell and tissue culture techniques. Predictable regeneration of plants with desired genotypes is of prime importance to successful application of crop improvement through molecular gene transfer. The cardinal factor to regeneration of agronomically important crops from cells rely in the technology to control the mutation and destruction rate of genes in culture stress cells. The general result of cell culture is somaclonal variation, which is not desirable when the engineered crop plants with uniform genotype and phenotype are needed. Scientists are addressing the molecular bases of variation to improve plant tissue culture techniques, but the results are a long way off. In the mean time, trial and error reigns.

Protection of plants from weeds, fungi, insects, and bacteria with pesticides and herbicides and other agricultural chemicals is a costly grower input and the challenge is to reduce cost of this agronomic practice. Pesticide use has been the most effective control of disease and pest control over the years. Overall, the use of pesticides to control disease infestations is rapidly changing because of the following reasons: 1) pesticide cost; 2) pesticide impact on the environment, that is humans, animals and other plants; 3) registration of pesticides and associated regulatory problems; 4) increased resistance of pest and pathogens to pesticides.

Genetics and breeding are the desired approaches to solving pest problems, many of which have been successful, but the results are generally slow to attain. It is evident that a lack of basic information about pathogens exist as well as host-pathogen interaction mechanisms. Part of the answers to the challenges before us can be met with tissue culture and molecular gene transfer technology coupled with continued breeding and genetic studies. Techniques involving *in vitro* culture methods that address disease problems are successful as tissue culture systems do as a selection system to determine resistance to fungal and bacterial diseases of plants.

Biological control may provide an opportunity to replace some pesticides for control of specific pests. In addition to the classical biocontrol techniques of collecting and release, the challenge is to develop new technologies to develop biocontrol agents. The use of genetic engineering and other manipulative techniques may be useful in transferring biocontrol molecules. Basic research on host pathogens and host insect interactions will also be needed to make this approach successful.

The above biotechnology impediments are the paucity of basic knowledge of plant biochemistry, physiology, and pathology and lack of uniformally effective DNA transfer technologies and reproducible tissue culture procedures. The most important of all of these is desirable gene identification, location in plant genomes and effective transfer of those genes. The ability to locate and isolate specific genes will lead to more efficient breeding for important agronomic traits e.g. drought and heat tolerance, pest resistance, reduced water contamination by pesticides and fertilizer, and an increase quality in yield. Until those genes are located, precisely characterized, and transferred, systematic and predictable production of plants with useful characteristics will continue to be developed through trial-and-error research.

Plant gene mapping research is the key to effective use of the new molecular biology techniques and realization of biotechnology potentials. Explicit knowledge about genes and rapid exploitation of that knowledge can anticipate or neutralize disease and problems with the interaction of harmful genotypes in the environment. With mapping information, new breeding schemes incorporating differently specified engineered gene complexes can be devised to help crop performance in each growing area. Plant genome research will also enable manipulation of genetic resources to achieve desirable variations. Scientist will be able to capitalize more effectively on natural pest resistance in plant; even where this resistance is initially weak, genetic manipulation can make it more effective. The benefits of such capabilities will be lower cost to farmers and consumers, and less chance of harming the environment with chemical pesticides.

The goal of the USDA Plant Genome Research Program is to find important genes and markers on chromosomes, isolate those genes from the genome, and transfer them rapidly to improve plant varieties. There is little knowledge of most important genes or of the regulation of their expressions. However, data from plant genome research will open the way to a concerted genome research program. The National Plant Genome Research Program will address single and multigenetic traits that relate to agriculture, forestry, and environmental concerns (i.e. global change and water quality). It will help maintain production stability and profitability and improve quality of agricultural products, maintain germplasms resources, and develop new crop resources. To accomplish these goals the National Plant Genome Research Program will foster and coordinate research. This research will lead to the ability to identify, characterize, alter, and rapidly and precisely manipulate genes controlling traits of agricultural importance to meet societies needs. Another goal is to establish a relational plant genomic database system to handle the mass of generated data. The plant genome research program information system will encompass a broader scope of genetic and physiological data than currently functioning databases. This broad goal will offer maximum use to the agricultural and scientific community. It will be able to use both the function of the molecular databases (e.g. GenBank, GenInfo, EMBL, PIR, MIPS, and others) and many concepts of designs and content definition of the specific-organism databases (e.g. E. coli genetic stock centered databases, *Coenorhabditis* system, mouse, *Drosophila*, and yeast centered databases and GDD). It must also expand and innovate in areas such as phenotype expression (including quantitative traits) biochemical pathways and physiological processes, diversity of karyotype and reproductive systems, and linkage to global species information. In summary then, it includes the world of purines, pyrimidines, amino acids,proteins, promoters and other molecular structures and plus many biological factors.

This is a formidable task to accomplish, even to make plans. A feasible approach, as suggested from the discussions of the plant genome database committee, ARS started with four prototypes agronomic species and *Arabidopsis thaliana*, a model plant. The scientists performing prototype data analysis are experimenting with different degrees of independent and coordinated development; with different amounts of overlap in areas or disciplines covered; and with different views of data modeling. Both success and failure are expected in the attempt to integrate crop-specific tailored components of the database.

There are hundreds, if not more, plants that are extremely important economically and scientifically. The program uses four agronomic species to represent some of the areas of diversity of genetic information and process. The ARS is also giving support to the National Science Foundations database efforts for the non-crop species *Arabidopsis thaliana*. The incorporation of more species would be desirable however, but the program has a limited budget. Also, there are other federal agencies and governments funding and/or supporting other crops such as rice in Japan. This number, however will sample diversity and the number is small enough to allow communication between groups as they choose to diverge or coordinate in the different approaches undertaken. The four species are corn, soybean, wheat, and pine. Corn is a commodity with a wealth of classic and molecular information on inbreeds, hybrids, experimental stocks, specific genes, and cytogenetic markers, and an extensive genetic map. Soybean also comes into play with a fairly good background in genetics breeding, some cytogenetics and other classical factors along with an increasing stable of information at the molecular level. Soybean is particularly important in terms of the economic aspects of gene systems that code for specific pathways and processes that regulate the generation of either fatty acids and lipids or oil production. Wheat is another economically important crop and it also has a well studied evolutionary component of allohexaploidy that is exploited in wide-crossing between species, resulting in alien substitution or addition lines. It also comes with excellent cytogenetic studies and active and cooperative international community. Tree species in general are an unusual choice for a genetic project. It was chosen for several reasons: the economic importance of the forest industry; the importance of biotechnology innovations to compensate the undesirable features of a long life cycle, limited research history, and an opportunity to facilitate and examine effectiveness of molecular mapping and other molecular biology approaches at an early stage; the importance of forest species in tying in with global environmental and ecological studies.

In the work to date we have come to speak of a 'genetic core' which was originally conceived as basic descriptors of the genotype of individual plants and plant populations and of the genome of the species. As the different cooperators have held 'users meetings' for the researchers in the crop community, the concept has been expanded to include another element that each group has defined for essential for initial utilization of the database, namely pedigree and germplasm information. The concept of the genetic core is evolving as the different groups are reaching a design phase which requires very specific decisions about how to represent the special elements needed to describe the genetics of their organism. One additional criterion for core data is 'do-ability' in terms of speedy implementation of critically important and immediately useful components. This defers such important components as handling quantitative traits and quantitative trait loci (QTLs) and relating pathways, phenotype, and physiology to implementation in later phases. This allows necessary time for innovative design during the first phase. The

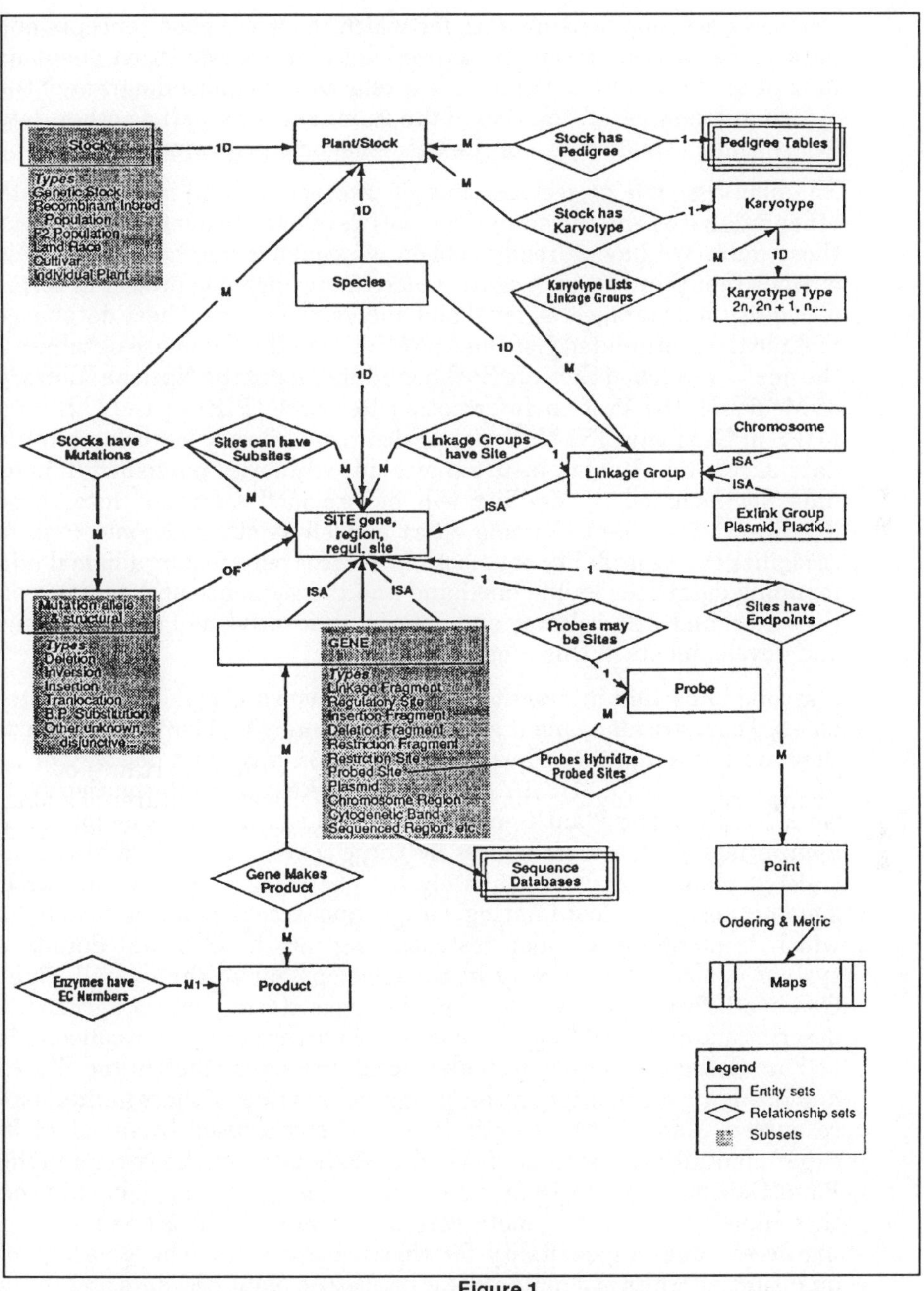

Figure 1

Overview of the genetic core of the plant genome database. The convictions of the extended entity relationship diagrams for relational databases are followed in general, with some specific adaptations. The rectangles show entity dash sets, the list under some of these entity sets are subtypes, which serve to illustrate what is defined by the entity name. The diamonds indicate relationship-sets and the cardinality of the relationship is indicated by using 1 or M on the arcs. Identity relationships are indicated by !ID!.

scientists are emphasizing data for which there are good concepts of a data model, access to well structured data, and a conviction about its immediate importance. Figure 1 is a relatively complex diagram of the systematic flow of information of the genetic core as well as other data items which would be part of the entire genome program.

The plant genome project anticipates interactions and interfaces with other databases and the analytical tools developed to query and analyze those data. We have already spoken of sequence databases, and it is obvious that they occupy a central place in any genome project in terms of acquisition, storage, retrieval and analysis software. These databases include the coordinated GenBank, EMBL, and the Japanese databases; the newly developed GenInfo Backbone database at the National Library of Medicine, the Protein Information Resource (PIR) in the U.S. and MIPS in Germany. The National Agricultural Library is working in the data acquisition level to insure that plant sequences published in journals not included in Medline will be provided for input into those databases. The Plant Genome effort as well as other genome centers recognize the desirability of smooth interfaces between organismal and mapping databases on the one hand and the sequence and analysis on the other, and will welcome and support innovative and creative ideas and developments in this area.

A second important interaction with germplasm and genetic stock databases. There are many plant genetic stock centers and for the most part, they are not computerized or computerized in a very ad hoc way or as direct tie-in with the USDA's Germplasms Resource Information Network (GRIN). The Plant Genome effort will encourage development of appropriate electronic databases for these centers as individual or centralized plans that will ultimately become integrated into the plant genome network. Coordinating design and access plans with GRIN, which is planning a major restructuring into a relational database system is also a top priority in the development in the overall Plant Genome Information System. This database stores general agronomic descriptions and is cataloged and is related to the germplasm collections at Fort Collins, Colorado and other locations across the United States and including much information on agronomic traits of these germplasm resources. Many of the genetic lines and stocks used by breeders in experimentalist are not included in the GRIN database, however and the Plant Genome Project will take a role in modeling and acquiring data on these lines as part of the genetic core, and work with GRIN on coordinating access and responsibility for these components. The coverage of quantitative traits is a topic of vital interest to plant breeders and there are many international groups establishing standards and procedures for exchanging information about these findings. Both the Plant Genome Project and GRIN must accommodate the fruits of these labors and be prepared to support and respond to these efforts.

The requirement to maintain broad germplasm resources (as well as evolutionary and environmental interest) make the developing taxonomic data bases an important partner in the Plant Genome effort. Currently, active efforts in the global Plant Species Information System and Species Planetarium encompass very ambitious and forward-looking database plans. Contact with this international effort facilitates mutual benefits and compatibilities for subject areas in which agricultural and taxonomic-evolutionary interest overlap. Modeling pathways, physiology, and phenotype (including quantitative and other agronomically important traits, is an area were the plant genome community will have to do much of the ground breaking. However, there are some related databases and database efforts. In the area of secondary metabolic product the NAPRALERT database of natural products literature, University of Illinois, at Chicago, available through the American chemical society, describes compounds, there pharmaceutical/biological activity, organism of origin and other factors. A number of efforts are underway to describe metabolic pathways and their regulation. Although the emphasis of the efforts as they develop may be on animals or microorganisms, the universality of much basic metabolism makes this work relevant to our initiative.

Summary

The advancement of biotechnology is hampered by lack of fundamental molecular knowledge in plant physiology and biochemistry, host-plant interactions, predictable cell and tissue culture procedures, and molecular DNA transfer methods that are applicable to all plants.

The greatest need to advance biotechnology is to precisely locate genes of desirable characteristics that code for traits such as disease resistance, environmental stress tolerance, yield increases and other enhancing factors. To accomplish this, the USDA Plant Genome Research Program is coordinating a national effort with a major goal of locating those genes in important agronomic and forest species.

The genome program will generate voluminous amounts of data that must be recorded, stored, and retrieved in an efficient and easily available electronic method for the scientific and commercial communities.

References

1) Plant Genome Conference Report. 1988. USDA S&E, ARS, Beltsville, Maryland, USA 20705.

2) USDA Plant Genome Mapping Program. 1989. Meeting Summary. Science and Technology Coordinating Committee. ARS, Beltsville, Maryland, USA 20705.

3) Data Resources and the Plant Genome Research Programs. 1990. A report by the subcommittee — Science and Technology

Coordinating Committee. National Agricultural Library, USDA, Beltsville, Maryland, USA 20705.

4) USDA Plant Genome Research. 1991. USDA, ARS - 94, Beltsville, Maryland, USA 20705

Sequence databases from EMBL and developments in electronic access

Peter J. Stoehr

European Molecular Biology Laboratory, Meyerhofstrasse 1, 6900 Heidelberg, Germany

Introduction

The explosion in data as a result of research in molecular biology has a major impact on research and development in biotechnology. Electronic access to up-to-date data is becoming increasingly important, and the requirement for comprehensive bioinformatics services is well-recognised [1,2]. Current important databases include information on nucleotide and protein sequences, structures of biological macromolecules, genome maps, genetic diseases, microbial strains, hybridoma, enzymes, cloning vectors, toxicological data, and abstracts from biological journals. It is desirable that much of the diverse activity in Europe be put on a firmer financial basis, with the resources to coordinate the research and development in bioinformatics necessary to allow Europe to compete as equals in international initiatives.

This paper describes the databases that EMBL Data Library currently distributes, and the developments in electronic access to them.

1 Databases available from EMBL

1.1 The EMBL Nucleotide Sequence Database

The nucleotide sequence database [3] was the original motivation for the establishment of the EMBL Data Library, and continues to be the main endeavour of the group. This work is done in collaboration with GenBankr [4] and the DNA Database of Japan.

Release 27 of the database (May 1991) contained 46,871 sequence entries consisting of 59.9 million bases. There is a rapidly accelerating rate of growth, a trend which will undoubtedly continue, especially as genome sequencing projects get underway. Clearly, the resources of the databases cannot increase at the same rate, and therefore data processing procedures have to be continually streamlined. The database is maintained in the ORACLE relational database management system for

```
ID   HSCA2   standard; RNA; PRI; 1523 BP.
XX
AC   Y00339;
XX
DT   19-SEP-1987 (Rel. 13; Last updated; Version 1)
DT   19-SEP-1987 (Rel. 13; Created)
XX
DE   Human mRNA for carbonic anhydrase II (EC 4.2.1.1)
XX
KW   CA2 gene; carbonic anhydrase II.
XX
OS   Homo sapiens (human)
OC   Eukaryota; Animalia; Metazoa; Chordata; Vertebrata; Mammalia;
OC   Theria; Eutheria; Primates; Haplorhini; Catarrhini; Hominidae.
XX
RN   [1] (bases 1-1523)
RA   David Hewett-Emmett Ph.D.;
RT   ;
RL   Submitted (01-JUL-1987) on tape to the EMBL Data Library by:
RL   David Hewett-Emmett Ph.D., University of Texas Health Science
RL   Center at Houston, Genetics Centers , graduate School of Biomedical
RL   Sciences, P.O. Box 20334, Houston, TX 77225, U.S.A..
XX
RN   [2]
RA   Montgomery J.C., Venta P.J., Tashian R.E., Hewett-Emmett D.;
RT   "Nucleotide sequence of human liver carbonic anhydrase II cDNA";
RL   Nucleic Acids Res. 15:4687-4687(1987).
XX
DR   SWISS-PROT; P00918; CAH2$HUMAN.
XX
CC   *source: tissue=liver; library=lambda gt11;
CC   *source: clone=pHCAII38.3 and pHCII14.1;
CC   **map: 8q22;
XX
FH   Key             Location/Qualifiers
FH
FT   CDS             39..818
FT                   /product="carbonic anhydrase II"
FT   polyA_signal    1079..1084
FT                   /note="polyA signal"
FT   polyA_signal    1266..1271
FT                   /note="alternative polyA signal"
FT   polyA_signal    1476..1481
FT                   /note="alternative polyA signal"
FT   polyA_signal    1506..1511
FT                   /note="alternative polyA signal"
XX
SQ   Sequence 1523 BP; 456 A; 309 C; 319 G; 439 T; 0 other;
gtgccgattc  ctgccctgcc  ccgaccgcca  gcgcgaccat  gtcccatcac  tgggggtacg
gcaaacacaa  cggacctgag  cactggcata  aggacttccc  cattgccaag  ggagagcgcc
agtcccctgt  tgacatcgac  actcatacag  ccaagtatga  cccttccctg  aagcccctgt
acagattgat  tcagtttcac  tttcactggg  gttcacttga  tggacaaggt  tcagagcata
......
atatatttat  agcaaagtta  tcttaaatat  gaattctgtt  gtaatttaat  gacttttgaa
ttacagagat  ataaatgaag  tattatctgt  aaaaattgtt  ataattagag  ttgtgataca
gagtatattt  ccattcagac  aatatatcat  aacttaataa  atattgtatt  ttagatatat
tctctaataa  aattcagaat  tct
//
```

Figure 1. A sample entry from the EMBL Nucleotide Sequence Database

efficient and flexible handling, and the need for direct electronic submission of data to the sequence databases continues to be critical in order to cope with the data flow: indeed, many journals now insist that nucleotide sequence data are deposited with the databases before a sequence-bearing article will be published.

Each database entry comprises a single contiguous sequence and its accompanying descriptive information (annotation). A sample entry is shown in Figure 1.

1.2 The SWISS-PROT protein sequence database

The SWISS-PROT database [5], maintained collaboratively by the EMBL Data Library and Dr Amos Bairoch (University of Geneva), is a collection of amino acid sequences translated from the EMBL Nucleotide Sequence Database, adapted from the Protein Identification Resource (PIR) [6] collection, obtained from the literature and directly submitted by research groups. SWISS-PROT is fully annotated and particular efforts are made to eliminate duplicate sequences and to annotate the presence and extent of sequence domains. Recent review articles are used to periodically update the annotation of families or groups of proteins. SWISS-PROT is rich in cross-references to other databases. SWISS-PROT is similar in format to the Nucleotide Sequence Database and therefore the two collections can easily be used together.

Release 17 of the database (May 1991) contains 20,772 sequence entries comprising 6.8 million amino acids.

1.3 Related databases

PROSITE [7] is a compilation of sites and patterns characteristic of specific biological functions found in protein sequences. It is maintained by Dr Amos Bairoch (University of Geneva). Some of the patterns have been published in the literature, but most have been developed by the database author. Cross-references are provided to instances of the patterns in the SWISS-PROT database.

ENZYME [8] is a database of characterised enzymes for which an Enzyme Commission (EC) number has been provided. It includes data such as the EC number, recommended and alternative names, catalytic activity and cofactors. Cross-references are provided to the SWISS-PROT database and also to the Mendelian Inheritance in Man database (MIM) [9] for human diseases associated with a deficiency of the enzyme. The main source of data is the recommendations of the Nomenclature Committee of the International Union of Biochemistry [10].

ECD [11] is a compilation of *E.coli* sequences in the EMBL/GenBank nucleotide sequence databases containing additional information on map locations. It is maintained by Dr Manfred Kroeger (University of Giessen). The CD-ROM distribution also includes query software for MS-DOS computer systems.

EPD [12] is a database of eukaryotic promoters, prepared by Philipp Bucher (presently at Stanford University, CA). This database contains detailed annotation of eukaryotic transcription start sites present in the Nucleotide Sequence Database and documented in the research literature. The database itself contains no sequences, but rather references to the sequences.

REBASE [13] is a database of restriction enzymes provided by Dr. Rich Roberts (Cold Spring Harbor Laboratory).

1.4 Capture of specialised biological annotation

The volume and complexity of sequence data makes it increasingly impractical for central data banks to maintain the full range of expertise required to annotate all sequences; indeed such annotation may be interpretive work of a kind more appropriate to specialised research groups. Databases maintained remote from, but coordinated with, a centralised sequence collection, allow the detailed biological annotation to be carried out at sites where the appropriate expertise is present. Figure 2 illustrates the current state of links between sequence-related databases, most of which are not centrally maintained. These links are manifested in database entries as pointers to stable objects in other databases, for example to primary accession numbers in the sequence databases.

2 Database access

Distribution of copies of the entire sequence databases is by mailing of CD- ROMs and magnetic tapes every three months. Much of this distribution is done by the Data Library and some is done by secondary distributors, such as groups which supply the data along with sequence analysis software. Further information about subscriptions can be obtained by contacting us at the address given at the end of this paper.

2.1 CD-ROM

CD-ROM is the preferred medium because it represents a cheap way to store large quantities data, and because the devices required to read it are within financial reach of the typical personal computer user. CD-ROM has quickly replaced magnetic tape as the most popular medium requested by users since its introduction in 1989 and we might expect magnetic tape distribution to be reduced still further if all computer manufacturers support the ISO 9660 format standard for CD-ROM.

The EMBL CD-ROM features:

- The EMBL Nucleotide Sequence Database and the SWISS-PROT Protein Sequence Database in the traditional ascii file format, with indices and documentation. This is a copy of the files distributed on magnetic tape.

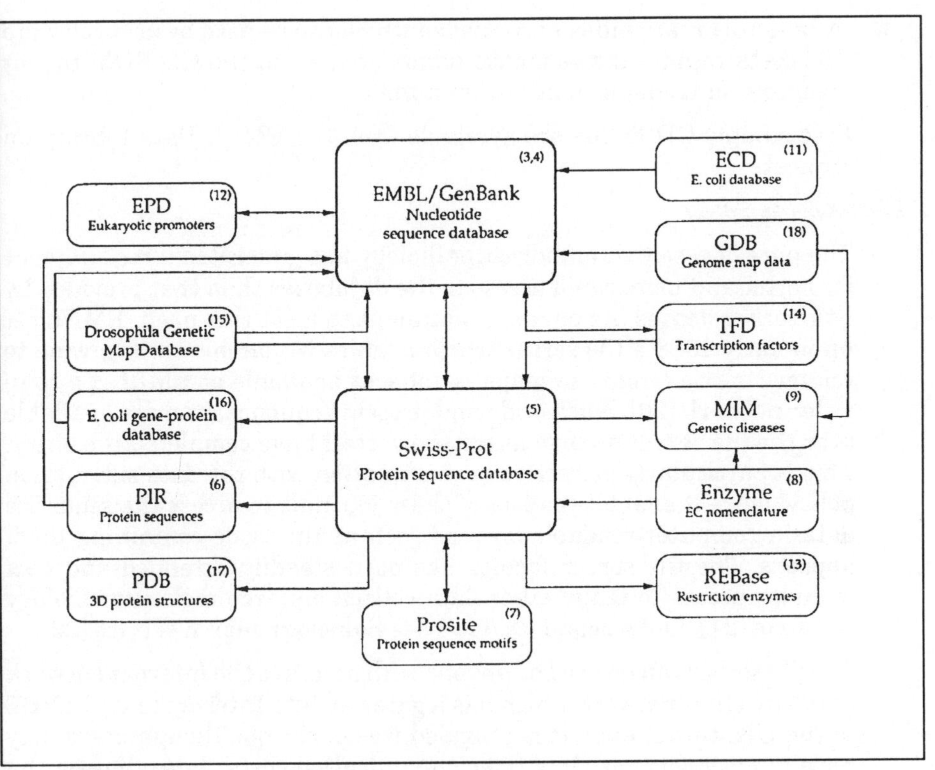

Figure 2

- Related databases: the Eukaryotic Promotor Database (EPD), *E.coli* database (ECD) with its own retrival software, restriction enzyme database (REBASE), PROSITE pattern database, and ENZYME nomenclature database.
- Query/retrieval software for MS-DOS systems, CD-SEQ, is produced in collaboration with Circle Information Systems B.V. (Heerlen, Netherlands) for query and retrieval of data from the two sequence databases. CD-SEQ is built on a commercial CD-ROM retrieval engine, FulText from Fulcrum Inc., and allows efficient and flexible retrieval of data using a range of text fields (e.g. entryname, accession number, literature citation, author, database cross-reference, free text and date).
- Sequence homology searching: The databases are also provided in a format compatible with widely-used sequence homology search software (e.g. FASTA by Lipman and Pearson [19] for MS-DOS and Macintosh systems). A program EMBLScan is also provided for MS-DOS systems for performing fast searches for very close similarity against the nucleotide database.

- A range of binary indices is provided which can be used by user software to access rapidly the sequence database files on the CD-ROM by, for example, entryname or accession number.

Free sample CD-ROMs are available from the EMBL Data Library on request.

2.2 Network File Server

The pace of research in molecular biology has generated a requirement for better and more rapid access to the databases than that provided by quarterly releases. As part of an attempt to meet this need, EMBL set up in early 1988 a file server which enables researchers world-wide to retrieve entries from the major databases available at EMBL via computer network [20]. Nucleotide and protein sequence data are available over the file server as soon as database staff have completed the entry. This is particularly attractive in combination with the data submission policy, since it enables readers of these journals to access the sequence data in computer-readable form as soon as the issue containing them appears. The file server facility has been steadily extended and now includes access to many other data collections, free molecular biology software [21], and a sequence database homology search service [22].

The file server can be used by anyone with access to the Internet network or to any other network which has a gateway into Internet (e.g., JANET in the UK, Bitnet etc.). It is provided free of charge, though users may have to meet some or all of the communication costs, depending on the accounting system of their local computer service. Use of the facility is simple and involves sending file server commands, one per line, in a standard electronic mail to the address NetServ@EMBL-Heidelberg.DE. The most important file server command, to get users started, is HELP. If the file server receives this command, it will return a help file to the sender, explaining in some detail how to use the facility.

2.3 On-line access via EMBnet

EMBL has striven over the past few years to establish remote copies of the nucleotide sequence database within Europe. This is designed to satisfy the demand for on-line interactive access to the sequence databases, and also to render them more available to the industrial community who are typically restricted from use of electronic mail networks to which academics usually have free access.

A collaborative project has been launched to establish a European molecular biology network (EMBNet) [23] consisting of centres of expertise in molecular biology and biocomputing in Europe. In 1988 a trial phase of the EMBnet project was initiated with four centres, and since then a gradual expansion has led to the involvement of a node in most western European countries, all of whom have been mandated, and some funded, at a governmental or research council level within their country.

National EMBnet Nodes

UK	Daresbury Laboratory, Warrington
France	CITI2, Paris
Netherlands	CAOS/CAMM Center, Nijmegen
Germany	DKFZ, Heidelberg
Spain	CNB, Madrid
Sweden	Biomedical Centre, Uppsala
Denmark	BioBase, Aarhus
Norway	Institute of Biotechnology, Oslo
Finland	Centre for Scientific Computing, Espoo
Switzerland	Biozentrum, Basel
Israel	Weizmann Institute, Rehovot
Greece	IMBB, Heraklion
Italy	University of Bari

Other nodes:

EMBL, Heidelberg, Germany
Hoffmann-La Roche, Basel, Switzerland
UK Human Genome Mapping Project, Harrow, UK
European Patent Office, Le Hague, Netherlands
ICGEB, Trieste
MIPS, Martinsried, Germany

Progress so far includes the establishment of network connections which have proved difficult across Europe due to the lack of infrastructure. Existing international networks are being used where possible (Internet, IXI, public packet-switching networks) with standardisation on the TCP/IP protocol. ISO/OSI networking standards will be phased in as appropriate. Systems to update remote copies of the nucleotide sequence database at national nodes on a daily basis have been implemented. These national centres make the data, along with analytical software, available to researchers, both industrial / commercial and academic, within their countries and offer training and support in their use.

Other network services are being investigated collaboratively (e.g. conferencing systems, remote access to specialised facilities) and original implementation is being broadened to embrace other types of nodes, particularly other database providers and hosts of specialised hardware or software that cannot realistically be duplicated at local sites.

Contacts are being made with other countries in Western Europe (e.g. Belgium and Ireland) and also Eastern Europe (e.g. Hungary, Russia).

The project has recently received a grant from the EC BRIDGE programme for the promotion of EMBnet network activities, as a result of a joint proposal coordinated by Professor Cecilia Saccone (University of Bari).

Future development

Advances in basic biological research and biotechnology are increasingly dependent on the effective management of information using computer technology. New approaches to the organisation of bioinformatics in Europe may be required to ensure that stable multidisciplinary expertise is focussed on it. One approach being explored is the establishment of a European Bioinformatics Institute (EBI) [24] to address the growth in database management and distribution activities, provide consultancy and training, to promote the development of new designs, tools and interfaces for relevant resources, enhance the professional and cooperative nature of the bioinformatics community, and to represent European interests.

How to contact the EMBL Data Library

Network: DataSubs@EMBL-Heidelberg.DE (for data submissions); DataLib@EMBL-Heidelberg.DE (general enquiries)

Postal address:EMBL Data Library, Postfach 10.2209, 6900 Heidelberg, Federal Republic of Germany

Telephone: +49-6221-387258

Telefax: +49-6221-387519

Telex: 461613 (embl d)

References

1. CEFIC (March 1990) Bio-informatics in Europe. 1 — Strategy for a European Biotechnology Information Infrastructure.
2. CEFIC (November 1990) Bio-informatics in Europe. 2 — Strategy for a European Biotechnology Information Infrastructure.
3. Stoehr, P.J. and Cameron, G.N. (1991) *Nucl. Acids Res.*, **19**, Supplement, 2227-2230.
4. Burks, C., Fickett, J.W., Goad, W.B., Kanehisha, M., Lewitter, F.I., Rindone, W.P., Swindell, C.D., Tung, C.S. and Bilofsky, H.S. (1985) *CABIOS*, 1, 225-233.
5. Bairoch, A. and Boeckmann, B. (1991) *Nucl. Acids Res.*, **19**, Supplement, 2247-2249.
6. George, D.G., Barker, W.C. and Hunt, L.T. (1986) *Nucl. Acids Res.*, **14**, 11-14.
7. Bairoch, A. (1991) *Nucl. Acids Res.* **19**, Supplement, 2241-2245.
8. Bairoch, A. (1990) University of Geneva, Geneva.
9. McKusick, V. (1990) Mendelian Inheritance in Man, John Hopkins University Press, Baltimore.

10. Enzyme Nomenclature, NC-IUB, Academic Press, New York (1984).

11. Kroeger, M., Wahl, R and Rice, P. (1991) *Nucl. Acids Res.*, **19**, Supplement, 2023-2043.

12. Bucher, P. and Trifonov, E.N. (1986) *Nucl. Acids Res.*, **14**, 10009-10026.

13. Roberts, R.J. (1985) *Nucl. Acids Res.*, **13**, r165-r200.

14. Ghosh, D. (1990) *Nucl. Acids Res.*, **18**, 1749-1756.

15. Ashburner, M. (1990) University of Cambridge, Cambridge.

16. VanBogelen, R.A., Hutton, M.E. and Neidhardt, F.C. (1990) *Electrophoresis* **11**, 1131-1166.

17. Bernstein, F.C., Koetzle, T.F., Williams, G.J.B., Meyer, E.F., Brice, M.D., Rodgers, J.R., Kennard, O., Shimanouchi, T. and Tasumi, M. (1977) *J. Mol. Biol.*, **112**, 535-542.

18. Pearson, P. (1991) *Nucl. Acids Res.* **19**, Supplement, 2237-2239.

19. Pearson, W.R. and Lipman, D.J. (1988) *Proc. Natl. Acad. Sci. USA*, **85**, 2444-2448

20. Stoehr, P. and Omond, R. (1989) *Nucl. Acids Res.* **17** (16), 6763-6764.

21. Fuchs, R. (1990) *CABIOS*, 6, 120-121.

22. Fuchs, R., Stoehr, P., Rice, P., Omond, R. and Cameron, G.C. (1990) *Nucl. Acids Res.* **18**, 4319-4323.

23. EMBL Data Library (1990) EMBnet: European Molecular Biology Network. EMBL Data Library Technical Document, EMBL, Heidelberg.

24. EMBL Data Library (1990) A Design for the European Bioinformatics Institute — EBI. EMBL, Heidelberg.

From spectra to structure: spectroscopic databases and the spectral analysis system SpecInfo

Michael G. Weller

Chemical Concepts P.O. Box 10 02 02, D-6940 Weinheim, Germany

Available Spectral Databases

Up until well into the 1980s, the bottleneck in large spectroscopic laboratories — as in Analytical Chemistry in general — was the acquisition of data (spectra), which was limited by cumbersome instrumental techniques and the need for sophisticated manipulation of the instruments during the recording of data. This bottleneck has been opened by the computerization of spectrometers (and other analytical instruments) and particularly by sampling automation and the introduction of robots.

In an economically-minded analytical laboratory, high-cost instrumentation such as high-field NMR spectrometers are operated 24 hours a day and 365 days a year. In most countries, however, this does not correspond to the working hours of analytical chemists.

Thus, interpretation of data became the new bottleneck. In spectroscopy, interpretation is aimed at determining the chemical structure of a compound based on the spectral data recorded. This process is based on rules and considerable experience. The efficiency of spectral interpretation was greatly increased by the availability of printed spectra collections which became available in the 1970s. Such hardcopy or microfiche reference works were quite effective in enhancing the efficiency of spectroscopic experts. Prime examples are the publications of Aldrich, Sadtler and VCH.

Name of database	Number of spectra (in thousands)							
	MS	IR	CNMR	XNMR[1]	Tape	CD-ROM / floppy	online (host)	Producer / distribution
Wiley Registry[5]	138				x	x	—	Wiley
MS Online	138				x	x	INKADAT	Wiley/FIZ Chemie
NIST	54				x	x	—	NIST
MSSS	42				—	—	CIS	NIST/CIS
SpecInfo	30	30	103	16	x	—	STN[2]	CC
CNMR/IR[6]		16	103		—	—	STN	Chemical Concepts
Sadtler IR Libraries		130			x	x	—	Sadtler
CSearch			50		x	—	—	Sadtler
SDBS	12	25	7		x	x	x[3]	NCLI[4]
Spectra Search				10	x	x	—	Fraser-Williams
Bruker Libraries		7	5		—	x	—	Bruker
Aldrich-Library		12			—	x	—	Nicolet
Sigma-Library		12			—	x	—	Nicolet
Nicolet vapour phase		9			—	x	—	Nicolet

[1] B-, N-, O-, F-, P-NMR [2] release: Dec. 1991
[3] Japan only

[4] Natl. Chem. Lab. for Industry, Japan [5] Contains a large fraction of the NIST spectra [6] largely identical with the SpecInfo database, will be discontinued when SpecInfo goes online

Table 1. Major Spectral databases

Since retrieval and comparison of data are of course much more effectively done by computers, the next logical step was the development of spectroscopic databases. Such databases are now available for a variety of spectroscopic methods, they contain a considerable number of spectra and are being used routinely in many laboratories, mainly for retrieval purposes and similarity searches.

Some of these databases contain spectra of rather inhomogenous origin and questionable quality. In some cases, especially in Japan, old versions of some of the databases mentioned below are distributed under different names. Further, some spectroscopic databases still do not contain chemical structures, or structures are stored as graphics only, i.e. without connections tables. This limits the use of the respective databases considerably.

Table 1 presents the larger and more widely used collections available today on magnetic tape, CD ROM, floppy disk or online, based on the information by respective database producers in March 1991.

Retrieval and similarity searches are obviously effective in the identification of known compounds. Accordingly, the US Environmental Protection Agency has commissioned a collection of standard MS data for environmental analysis studies (these are part of the Wiley and NIST databases).

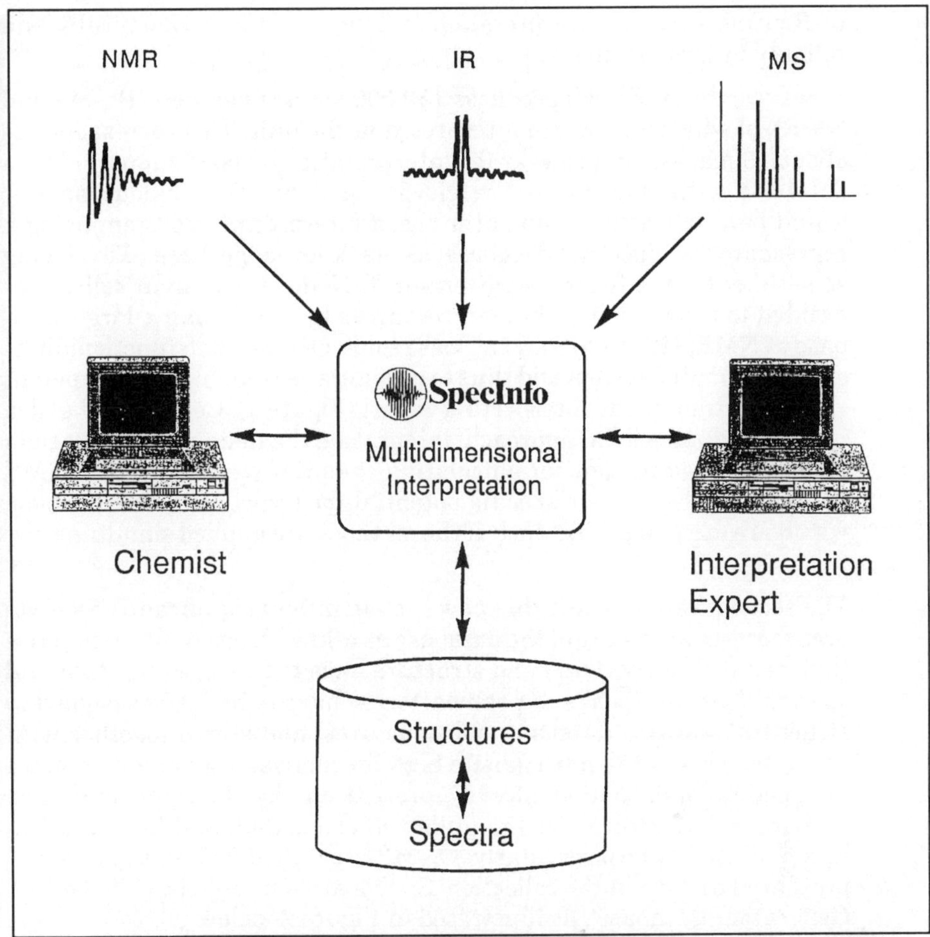

Figure 1. Spectroscopic information from MS, IR and a variety of NMR methods are analysed on the basis of a database containing correlated spectral and structural patterns

The SpecInfo Approach

Research of course mostly yields *new, unknown* compounds, the spectra of which by definition cannot be represented in the database. Spectral databases searched with retrieval software only are helpful, albeit of limited effectiveness in such an environment. This was recognized by W. Bremser at the spectroscopic unit of BASF's research lab in Ludwigshafen in the early 70s, when new automated sampling techniques and spectrometer computerization increased the turnover of spectra dramatically and led to the above-mentioned bottleneck in spectral interpretation. In this BASF laboratory's spectroscopic unit, spectra are routinely interpreted, i.e. chemical structures are determined, for two reasons. i) Research chemists are generally not primarily interested in the spectra, but in the structures of compounds they have synthesized.

ii) Routine spectral interpretation is done far more economically and reliably by 'professionals'.

However, the BASF lab produces 150 000 spectra per year (IR, MS and NMR), of which 70 000 are interpreted in the unit. This corresponds to about 10 man-years per year for interpretation, even if supported by a database with conventional retrieval software. The obvious answer would be an 'automatic' computer-based interpretation system, using a representative spectral database as its knowledge base. This being something for the future, W. Bremser, R Neudert and their colleagues decided to save valuable human resources by developing a large database of NMR, IR and MS and at least some modules and functionalities of an 'automatic' system and thus to provide a reasonable and competent 'dialog partner' to the interpreting expert (Figure 1). Central to the idea is a multi-dimensional approach, integrating the main methods used in molecular spectroscopy for elucidating chemical structures. MS, NMR and IR each have their specific potential for structure determination, which are used most efficiently if the methods are applied simultaneously.

The second way in which this new system differs significantly from its predecessors, is the use of the database as a 'knowledge base'. In practice this means that spectrum and structure have to be stored together and spectral features (peaks or peak patterns) have to be made assigned to structural features (atoms or substructures) and stored together with the data. This is the prerequisite both for analysing unknown spectra for spectral information already present in the database and thus deriving substructures for the unknown compound, and for the calculation of spectra (particularly CNMR) for chemical structures not presented as such in the collection. Such a system could be called a '2nd Generation Database', as illustrated in Figure 2 below.

Spectral Analysis

A variety of programs then make use of this 'knowledge' enabling the user to go both from spectrum to structure and from structure to spectrum when doing a semi-automatic interpretation of spectral data. A large variety of procedures are available; two typical-ones are described here:

1. Analysis of spectra (from spectra to structure)

- Input of query spectrum, e.g. CNMR, recorded of unknown compound
- Analysis for spectral features
- Search for such features in the database
- Generation of substructures related to these spectral features
- Statistical analysis for the relevance of substructures and ranking of hit-list
- Production of hitlist of structures and/or substructures as candidate structures for the unknown on the basis of the CNMR characteristics

Database of 1. Generation

step 1 **Cards / spectra atlas**
access only via sort-criterion
step 2 **catchwords**
access via subject-index too
step 3 **logical connection of hitlists**
concentration of information via
intersection of hitlists
step 4 **substructure- and full-text-searches**

● **Target Information of Retrieval:**
any information that is
EXPLICITLY
part of the library (e.g. existing publications,
properties of existing compounds)

Database of 2. Generation

● **Target Information of Retrieval:**
prediction and estimation of properties of
compounds that are
NOT
part of the library

Figure 2. The '2nd Generation Database'. What is outlined here would often be addressed as an 'expert system'. On purpose this elusive term is not used here. However, SpecInfo is aimed at being 'the expert's system', indicating that it competently supports the spectroscopic expert in analysing spectral data.

- Application of basically the same procedure to other spectra of the same compound, if required, such as IR, MS, or hetero-atom NMR, again producing hitlists of (sub)structures derived from the respective methods
- Combination of hitlists and/or generation of complete candidate structures for the unknown using a structure generator
- Critical examination and correction of the hitlists produced, by the interpretation expert, applying experience, common sense and information available, e.g. from the synthetic chemist
- Simulation of spectra for structures in the hitlist and comparison with the query spectrum, thus reducing the solution space and producing a final ranking of candidate structures for the unknown compound.

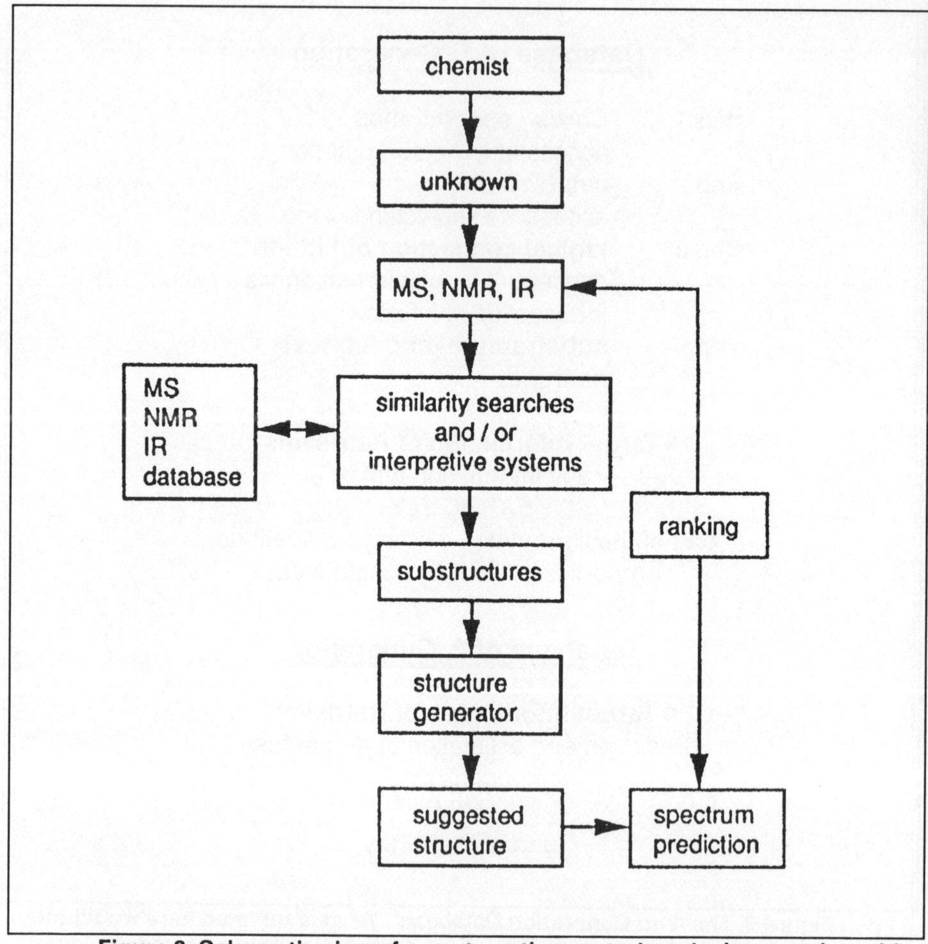

Figure 3. Schematic view of an automatic spectral analysis operation with SpecInfo. Courtesy of R. Neudert, BASF.

This procedure is also illustrated in Figure 3. An example for a combined CNMR and IR application is given in Figures 4a-c below.

2.Spectrum calculation (from structure to spectrum)

In many cases the synthetic chemist can produce a very educated guess indeed as to the structure of the compound produced. It may then be sufficient to calculate the CNMR spectrum for the proposed chemical structure and compare the result with the spectrum recorded to confirm the structure. An example of a structure calculation is given in Figure 5 below.

Using these and other operations of SpecInfo, BASF has cut down manpower for interpretation (see above) of 70 000 spectra per year from 10 to 5 man-years per year.

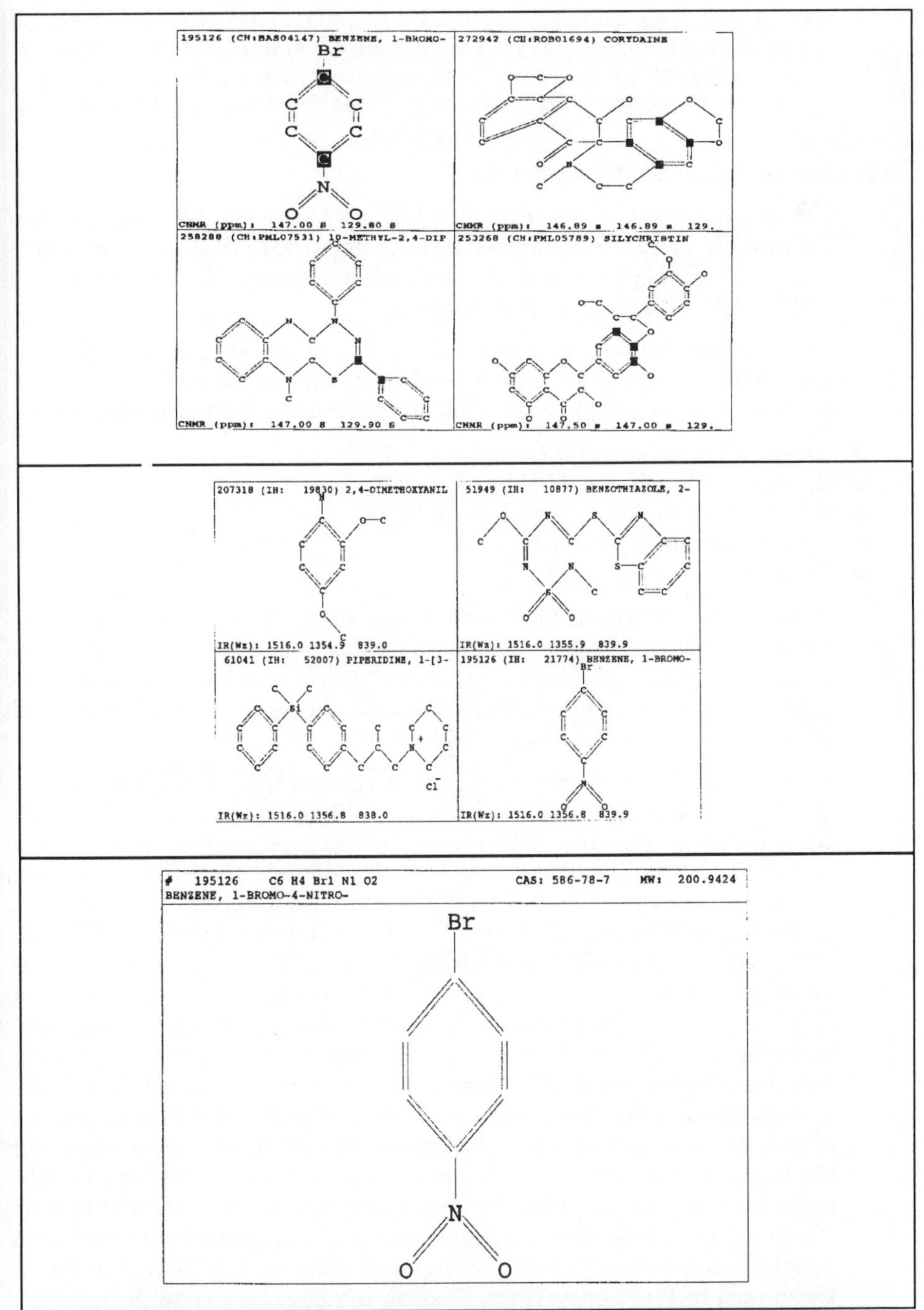

Figure 4. Combined C-NMR and IR search: to identify an unknown compound C-NMR and IR spectra were recorded. 4a. A line search for two singlets with chemical shifts at 147 and 129.8 ppm results in a hitlist of 150 structures, of which the first 4 are shown. 4b. Similarly, a search for the

continued over page

resonances at wavenumbers 1516, 1356 and 839 results in another hitlist of 22 structures of which the first 4 are shown. 4c. At this point the substance can be identifed by logically coupling the two lists, an operation which yields the displayed structure as the only one satisfying both the NMR and IR criteria.

SpecInfo — a co-operative development

The development of SpecInfo at BASF was sponsored by the German Federal Ministry of Research and Technology (BMFT). As a result of this grant, the database was made available online by FIZ Karlsruhe and STN as CNMR/IR online (see Table 1).

The German government had passed a program supporting scientific information systems in general and those on the field of chemistry in particular. This program includes five chemical information systems:

- Organic Chemistry (Beilstein)
- Inorganic Chemistry (Gmelin)
- Chemical Reactions (ChemInform RX)
- Thermochemistry (Detherm)
- Spectroscopy (SpecInfo)

In 1989, Chemical Concepts proposed the integration of several spectroscopic software and/or database activities towards a integrated spectroscopic information system. Rather remarkably, the BASF agreed to make publicly available about half of its internal spectroscopic database and the entire software developed for SpecInfo.

A number of spectral collections and programs was secured to complement the BASF development: the Max-Planck-Institute in Mülheim contributes its MassLib software for the interpretation of mass spectra and GC/MS runs as well as its high-quality MS database. The simulation of mass spectra is being developed by Prof. Gasteiger (TU Munich). A B-NMR database compiled from literature is contributed by Prof. Nth (University of Munich). Chemometric aspects are taken care of by an exploratory data analysis system (EDAS) developed by Prof. Varmuza (TU Vienna). The Institute for Spectrochemistry and Applied Spectroscopy (ISAS, Dortmund) agreed to contribute its developments in quality assessment of spectral data. Soon it became obvious that a fair number of organisations felt the need to join efforts devoted to different aspects of spectroscopic software and databases into one integrated system. On the basis of this policy, the German research ministry decided to substantially support this joint effort, which is now being coordinated by Chemical Concepts. Major enhancements to the group contributing to SpecInfo, come from Sumitomo Chemical (Japan), ICI (UK), Toyohashi University of Technology (Prof. Sasaki), the ETH Zrich and the German Cancer Research Center. Figure 6 below provides an overview of software modules presently being integrated into and developed for the SpecInfo system.

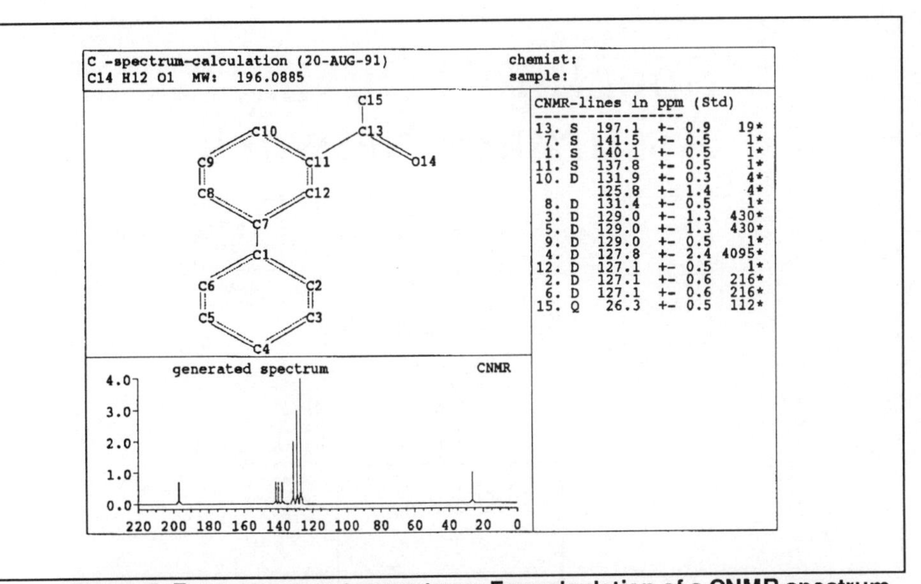

Figure 5. From structure to spectrum: For calculation of a CNMR spectrum, the query structure is entered with SpecInfo's graphical structure editor or with the MDL structure editor (e.g. ChemBase). The display shows the estimated spectrum of 3-phenylacetophenone accompanied by a list of the calculated signals with tolerances from which the reliability of simulation may be judged.

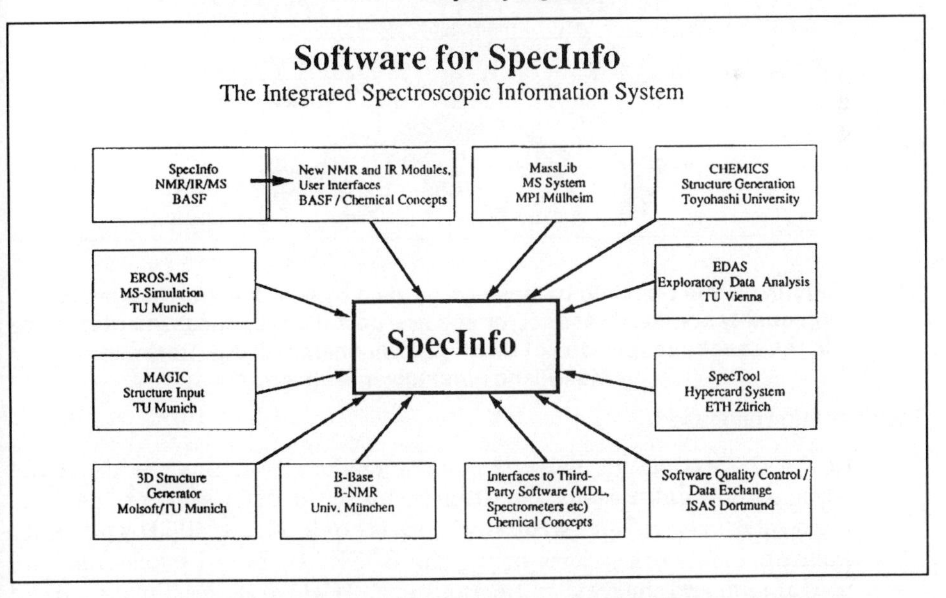

Figure 6. A remarkable group of organisations, both industrial and academic, contributes or develops software modules for SpecInfo together with Chemical Concepts.

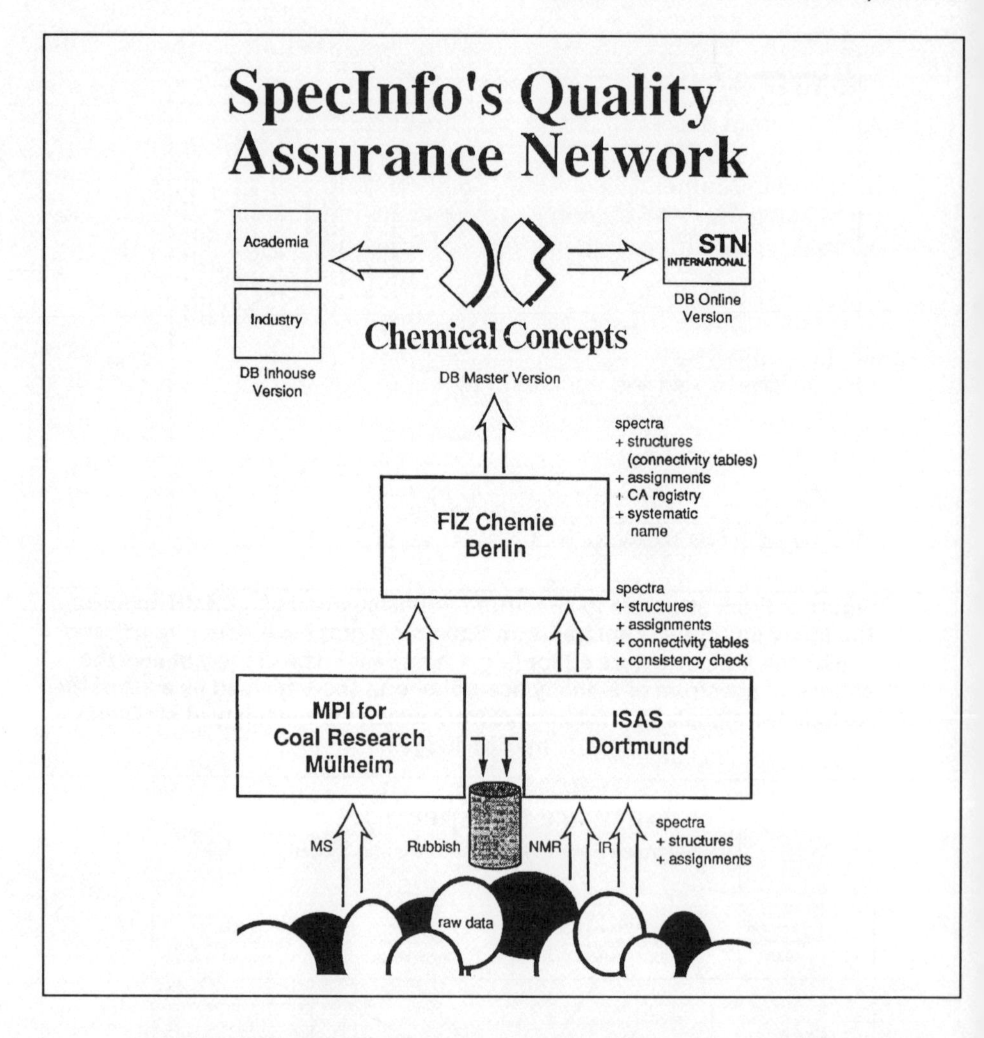

Figure 7. The SpecInfo datbase being used by way of a knowledge base, high quality standards are set for spectral data. Quality assessment is made in the spectroscopic labs of BASF (Ludwigshafen), ISAS (Dortmund) and Max-Planck-Institute in Mülheim.

The SpecInfo Database

In the context of the above it is quite obvious that the quality of the spectroscopic database is a limiting factor to the usefulness of SpecInfo. Each set of spectral and structural data is quality-controlled by a quality assurance network incorporating the BASF, the Max-Planck-Institute in Mülheim and the ISAS in Dortmund. FIZ CHEMIE, Berlin has agreed to contribute systematic data such as chemical abstracts registry numbers, systematic names and also to check the consistency of all systematic data. Figure 7 above depicts the system established for quality assurance.

An important aspect of quality assessment is the assignment of quality vectors to the Wiley MS database carried out by the Max-Planck-Institute in Mülheim. The large Wiley collection is thus being split into three categories — full, reduced and unsatisfactory spectra — and will be used for qualified searches within the SpecInfo system.

It was estimated, that a database representing spectral data of all relevant methods with statistically relevant information on all existing chemical substructures would have to consist of at least 100,000 spectra each of MS, IR, CNMR and HNMR, and at least 10,000 spectra each of B-, N-, O-, F-, P-NMR and Raman. Thus, a minimum of 500,000 spectra will be required.

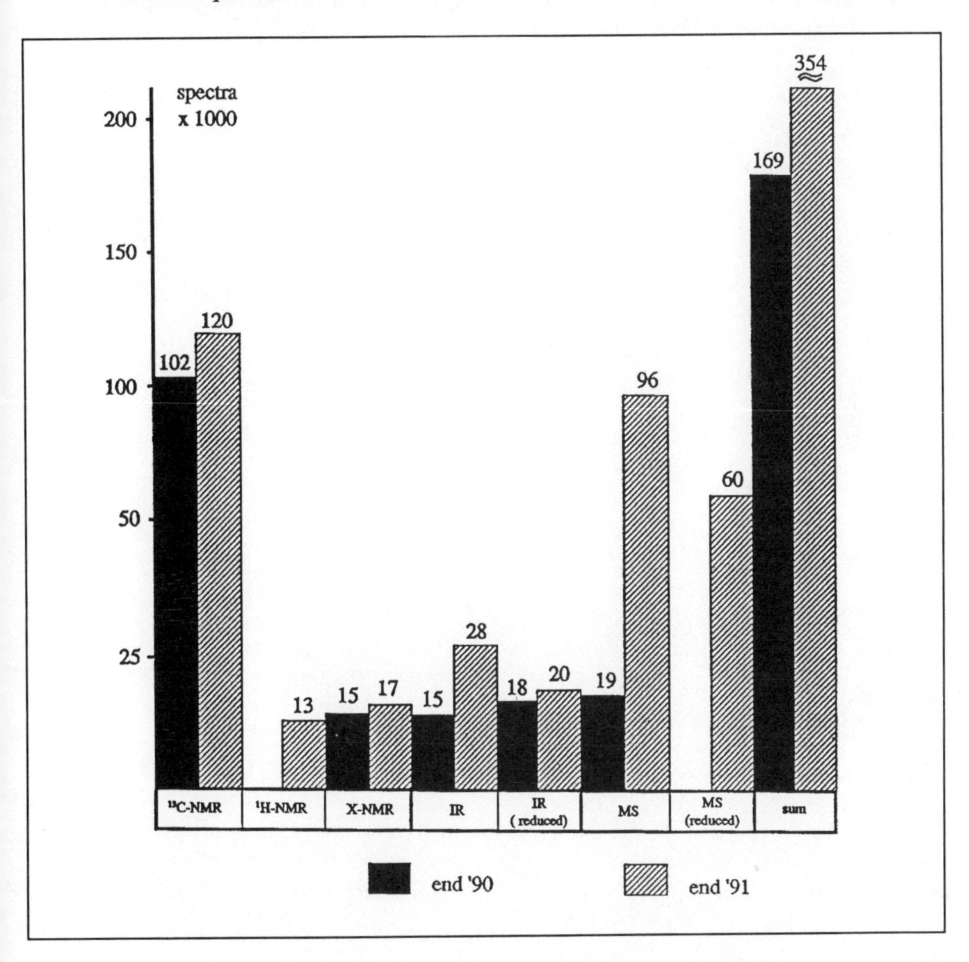

Figure 8. Status and growth of the SpecInfo database.

A considerable number of organisations have agreed to feed their data into the SpecInfo spectral database. As a result of this, the target of at least 500,000 spectra could be met in the near future. The present status and anticipated growth of the SpecInfo database is given in Figure 8. above.

Availability

Specinfo is presently available on tape for inhouse use. With this inhouse version, users can file their own data into the system, thus enhancing the knowledge base specifically in the area of the user's chemical activities.

The online database CNMR/IR online (see Table 1) will be replaced by SpecInfo online (designed to incorporate the full set of programs and data), which is currently being implemented under Messenger on STN for release in December 1991.

Flexible fitting of target molecules to 3D pharmacophore models

C. M. Venkatachalam, Ryszard Czerminski and Rudolph Potenzone, Jr.

Polygen Corporation, 200 Fifth Avenue, Waltham, MA 02254.

I. Introduction

One of the difficult tasks in designing new pharmaceuticals is the understanding of the requirements for binding to the biological receptors that will elicit the desired response. While there are many other factors that are important in the clinical efficacy of a drug, clearly the binding event is essential in the process.

The focus of our work has been in the development of techniques that can be used to assess the quality of 'fit' between a target drug and the assumed pharmacophore model. There are three separate aspects to this problem:

1. Identify or guess the receptor/pharmacophore model. In so doing, we must be able to define a metric that describes the 'fit'.
2. Optimize the shape of the target drug to find if it can 'fit' (i.e. fall within the 'fit' criterion)
3. Evaluate the 'fit' with other criteria such as internal energy of the target molecule and accept or reject the 'fit'

II. Pharmacophore models

It can be difficult (or impossible) to identify a specific pharmacophore model that has predictive capability for a particular class of activity. Use of traditional rational design and computer based techniques can often lead to insight for the discovery scientists to develop a pharmacophore model. Inspection of the activities of derivatives (such as Marshall's Active Analog Approach [1]), results from QSAR or CoMFA [2] or analysis of simulation results (see the proposed sweetness receptor [3]) can all lead to the development of aspects of a pharmacophore model. Automated procedures to try to identify what the pharmacophore may be have not been successful.

One type of receptor model can be viewed as identification of regions in space that have particular requirements of themselves as well as requirements with the other regions. The types of regions include areas of positive or negative charge, hydrogen bonding regions (acceptors or donors), and regions of exclusion (no atoms can occur in this region). The relationship between regions is generally on the basis of distance or angle relationships but these can be complex (certain distance along a bond vector).

These regions of space can be identified by intuition on examination of the compound data. Alternatively, a number of field analysis algorithms have been used to examine similarity of molecules based on how they are perceived under certain conditions. Electrostatic field or steric field analysis has been used for these purposes for many years by Smith [4], Cramer [5] and Richards [6]. From the information on the various calculated fields, it is possible to identify the key regions for further analysis.

One problem with this type of receptor model is in the evaluation of the 'fit' of a target molecule to the various desired properties. Without a specific metric that describes the goodness of fit to the model, it is impossible to design automated algorithms to seek optimal fits. However, they have the excellent advantage of stimulating intuitive drug design by discovery chemists.

An alternative method for defining a receptor model is to define atom matches with respect to an active compound. However, rather than taking the entire active molecule, only specify the parts of the active compound that are essential for binding. This is quite similar to Marshall's approach. The difficulty is often trying to identify the active conformation of any of the highly active compounds. Many methods have been attempted to do this, quite often involving exhaustive search techniques on a subset of compounds that are the most rigid. These are then analyzed to try to find a particular conformations which all of the active compounds can all adopt but may be energetically less stable for the inactive compounds.

An alternative approach for identifying the active conformation is the Molecular Shape Analysis (MSA) of Hopfinger [7]. This method allows all stable conformations of all of the compounds in the study to be used serially as the active conformation. Through 3D-QSAR analysis using 2D and 3D descriptors for thermodynamic, structural and electronic properties, the conformation that best fits the data is identified.

III. Flexible fitting

In the case of atom matching, it is easy to define the quality of fit by simply the root mean square (RMS) of the distances between the pharmacophore model and the target molecule. Further, it is possible to use the RMS as the objective function in a simplex search scheme of torsional angle rotations to optimize the quality of the fit.

We begin our method with a rapid rigid body superposition which is completely stable. We independently derived this approach but it has been previously reported by Kearsley [8]. At every step in the simplex minimization, the target molecule is rotated and translated according to the superposition scheme.

All rotatable bonds are allowed to vary during the process and this converges very quickly depending on the number of rotatable bonds to a minimum in RMS.

Since this algorithm is dealing with rotatable bonds that lie on a complex hyperspace, any solution to minima cannot be assumed to be at the global minimum. We have studied the effect of random starting points on the converged structure and found that as the flexibility of the target molecule increases, there is a dramatic increase of local minima. This must be considered in any study. We have also examined the possibility of starting from multiple start points for our target structure based on the conformational flexibility and conformational studies prior to the flexible fitting.

It is also possible to add to the fitting process other variables along with the RMS. This is done by creating a complex objective function for the simplex optimizer such that:

$$\Omega\ [\mathbf{Molecule}^{target}] = \mathbf{RMS{+}A{\cdot}Energy}$$

The arbitrary scaling constant A serves to adjust the relative weights of RMS versus the Energy in the optimization. As A goes to 0, then this becomes a simple optimization of the RMS fit between the target and the model. As A gets large, RMS is ignored and this become a simplex minimization of energy. We have experimented with A and found that it is possible to initially have A set to 0 and then as convergence is achieved, allow it to rise slowly. The next effect is that initially the target molecule fits well to the model and, depending on the energetics, it will move away from the good fit to what is the best overall fit within the energy tolerance.

IV. Cartesian flexible fitting (template forcing)

An alternative approach to fitting a molecule onto a proposed model is to perform a rough Cartesian minimization including target constraints for the atoms that match to the model. Thus, the energy function for the calculation is modified such that the difference between the atomic coordinates of the target model from the atoms in the model is a severe penalty function. During the simulation, the energy computed is only the internal energy of the target molecule plus the constraints.

$$\textbf{Energy}_{Total} = \sum_{Atoms} \textbf{E}_{Internal} + \sum_{n} \textbf{E}_{Constraints}$$

The constraints are applied with a simple quadratic function, one for each matching atom in the target molecule to each point in the receptor or pharmacophore model.

With this new energy function so defined, the target molecule is subjected to a full Cartesian minimization regime (conjugate gradient for 200 steps). The net effect of this approach is first to (sometimes) brutally force the molecule into the shape of the model, regardless of the internal energy. This can give intermediate states that have very poor bond lengths and bond angles as well as overlapping atoms. However, as the minimization achieves the constraints of the atoms onto their desired positions which effectively gets the energy of the constraints to zero, the molecule then minimizes to a close local minimum. Since the summation remains in effect, the resulting minima will be a high energy structure, stabilized by the constraint energy. The final step of this approach is to completely turn off the constraint energy term and allow the minimization to proceed. This yields an energetically stable conformation as near to the pharmacophore model as the energetics will allow.

The problem with this method is that it can be too rough with the starting structure, even to the point of inverting chirality or shifting too far from a good minimum energy configuration. However, when the flexible fitting approach is not yielding good results, this can provide a very important approach.

V. Results

The test system that we have studied is two bi-cyclohexyl structures where the receptor model is defined by the two rings separated by five bonds and the target molecule has the two rings separated by twelve bonds. This system represents a case where there are multiple solutions, probably beyond the range where any systematic study is possible. An appealing feature of this particular problem is that the mapping from the model to the target is obvious since it is clear that the two rings in the target need to be superimposed onto the two rings in the model. Further, the RMS fit between the two rings could be perfect while the many bonds between the rings find their low energy conformations.

The RMS is defined as the six atoms in each ring matched to the six equivalent atoms in the matching ring so that there are twelve distances in all. The target structure was energy minimized and the lowest energy conformation was taken as the fixed model. The flexible molecule was set in a fully extended structure (all rotatable non-ring bonds set to 180 degrees). This was then energy minimized and the resulting conformation was very close to the extended structure. Simple superposition results in a very bad match between the two structures and an RMS of

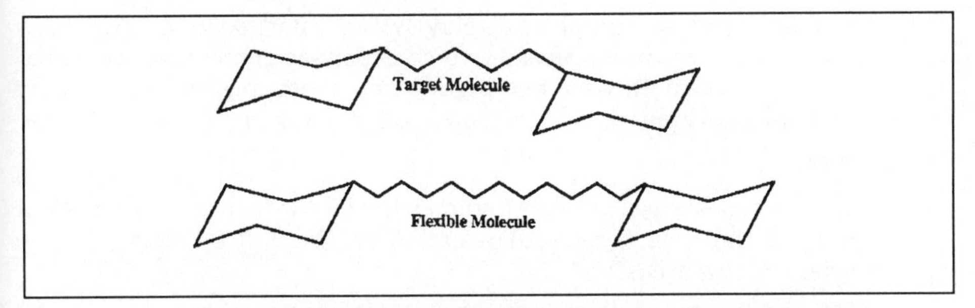

Figure 1. The target and flexible molecules used in the study

4.89. Allowing the twelve rotatable bond to become flexible in the target molecule, the RMS was able to be reduced to 0.50 after only 300 steps of Simplex optimization. The internal energy (CHARMm [9]) was not minimized in any way (i.e. A was set to 0) and was very high. As A was increased to .01 and the fitting carried out again, the energy of the target molecule dropped to 47.3 k cal/mol as the torsion angles were minimized to lower barriers for the dihedral energy terms. RMS fit increased slightly but then moved to 0.38 since there is enough flexibility to reduce the energy while maintaining a good RMS fit.

Cartesian flexible fitting was used and, after 500 steps of minimization, the RMS fit was perfect at 0.0 while the energy was reduced to 23 K cal/mol. In this particular case, this method was quite reasonable and yields better results. There is enough flexibility in the 12 bonds to adopt a perfect match while still minimizing the energy.

Finally, 500 random conformations about the 12 bonds were generated to product random starting structures for the flexible fitting. All 500 were allowed to fit onto the model with no energy (A=0) and then the energies were computed during the fitting. The 15 lowest energy structures were evaluated and all were in the RMS range of 06 and 0.9. The internal energy of the 15 structures ranged from 200 kcal/mol to very high energies. However, by inspection, there was quite a diverse set of shapes for the atoms of the 12 connecting bonds. This approach, however, proved very robust and very fast. Each of the 15 structures was allowed to relax the chain atoms (connecting non-ring atoms) while holding the rings fixed in space. The resulting energy range was 4.9 to 83.8 k cal/mol while holding the exact RMS values.

The last step in this process, we combined the 10 lowest energy structures into a single volume map examining the intersection volume which showed that only the rings were fully intersecting and that the connecting chains spanned a wide range of space. However, we also made a volume density map. The overall volume contained by the structures was divided into a grid of points and each point assigned a density value equal to the number of molecules that contained that point. Each point was then displayed on a graphics display system color coded by the density value. This density map showed which regions in space were covered by any molecule in the set but highlighted those regions which were maximally covered.

VI. Future work

The efforts described here are a framework for the future research of our group in the area of molecular similarity. We will be moving into a variety of new areas:

1. Field fitting. We will be incorporating the ability to measure the 'fit' of a molecule to a calculated field. We are experimenting with techniques for combining the results of grid maps (such as Peter Goodford's GRID [10] or the QUANTA Probe calculations) where there are equally spaced grids of points which each contain a computed value. From these values, iso-value surfaces are constructed. The approach that we are experimenting with is to compute similar maps of the target molecule and try to maximize the grid points on the surface that match up. Alternative methods have been published by Smith [11], Hermann [12]and van Geerestein [13].
2. Once we have defined a mechanism to fit the fields of two molecules and thereby compute an index of similarity, we can in an equivalent manner use this index to perform flexible fitting. In this way, molecules can be forced to match the field of a receptor model.
3. We also expect to be able to similarly treat the volume density maps that come from the flexible fitting algorithms. Once a volume density is computed and the regions of high density are selected, it should be possible to flexibly fit a molecule into the maximum density regions, without worrying about the atom matching.

VII. Conclusion

We have described a mechanism for performing flexible fitting of molecules onto templates. What is required is the ability to measure the quality fit at the current stage. We have experimented with a very fast Simplex approach which has proven to be both fast and effective for flexible fitting of one compound onto another using RMS atom superposition.

References:

1. G. R. Marshall, C. D. Barry, H. Bosshard, R. A. Dammkoehler and D. A. Dunn, 'The Conformational Parameter in Drug Design: The Active Analog Approachs', in Computer Assisted Drug Design, American Chemical Society Symposium Monograph Series No. 112, Washington DC (1979) pp. 205-226.
2. R.D. Cramer, D. E. Patterson and J.D. Bunce, *JACS,* **110** (1988) 5959.

3. J. C. Culberson and D. E. Walters, '3-D Model for the Sweet Taste Receptors in Sweeteners: Discovery, Molecular Design and Chemoreception', edited by D. E. Walters and J. Brady, American Chemical Society, Washington DC (1991) pp 214 to 223.

4. G. Smith, SEA4: Quantum Chemistry Program Exchange, Bloomington, IN.

5. R.D. Cramer, D. E. Patterson and J.D. Bunce, *JACS*, **110** (1988) 5959.

6. G. Richards, ASP software available from Oxford Molecular.

7. M. Mabilia, R. Pearlstein and A.J. Hopfinger, in 'Molecular Graphics and Drug Design', edited by A.S.V. Burgen, G.C.K. Roberts and M., S. Tute, Elsevier, Amsterdam (1986) p158.

8. S. Kearsley, *Acto Cryst.*, **A45**, (1989) 208-210.

9. B. Brooks, R. Bruccoleri, B. Olafson, D. States, S. Swaminathan and M.J. Karplus, *J. Comput. Chem.*, **4**, (1983) 187.

10. P. Goodford, *J. Med. Chem.*, **28**, (1985) 849.

11. G. Smith, SEA4: Quantum Chemistry Program Exchange, Bloomington, IN.

12. R. Hermann, SUPER program, paper submitted for publication.

13. V.J. van Geerestein, P.D.J. Brootenhuis, C.A.B. Haasnoot, presented at the American Chemical Society National Meeting in New York, Aug. 1991.

Patent databases: a critical overview of databases and media

Irène Savignon

INPI, 26bis, rue de Léningrad, 75800 Paris Cedex 08, France

Before doing an overview of patent databases or instead of doing it, a first question could be: why are there so many patent databases? Why patents? What kind of relationship is there between inventions and humans? What are the reactions facing inventions?

Inventions disturb

New creeds, new ideas, new discoveries and new inventions tend to disturb the established order of things, sometimes to the point of utterly destroying them. Accordingly, every new creed, idea or invention is bound to meet the opposition of the supporters of the established order. Workers often oppose new inventions in order to preserve their jobs and avoid unemployment. Many cases of riots leading to the murder of inventors and the destruction of new machines are recorded.

Competitors of the inventor are generally adverse to novelties able to upset their production methods and their commercial markets. At the time when production and trade were organised by guilds and corporations, introducing new techniques without the consent of the guild was deemed equivalent to unfair competition.

And more than once, the rulers of the past supported these points of view. The Emperor Vespasian, according to Suetonius, rewarded the inventor of a labour-saving device but refused to have it used in practice. The same reasoning justified the rejection of applications for patents in seventeenth century France; then the notion of 'usefulness' was primary for granting a patent, and, at a time when unemployed vagrants were an aching sore for the government, less work for men was looked upon as anything but useful.

Extreme care for the values of order and stability is to be found in Japan under the rule of the Tokugawa Shoguns. Not only did they prohibit nearly all kind of trade and exchange with foreign countries, but they also banned native innovations of every description. The law establishing this prohibition stayed in force for about one hundred and fifty years, and was abrogated only in 1868, at the beginning of the Meiji era. Two years later, the first Japanese patent law was passed, but nobody filed an application (Japan is today by far the leading country for the number of patents filed).

Inventions, the stuff of progress

Most governemts indeed took a view completely opposite to the policy of the Tokugawa Shoguns. Especially in the seventeenth and eighteenth centuries, with the development of the idea of progress, new inventions, together with new books and new works of art, came to be looked upon as the very stuff of enlightenment, at least in western countries. Even if the first known patent law is to be credited to the Republic of Venice as early as 1474, it is the celebrated English 'Statute of Monopolies' of 1624 that is considered the cornerstone of the patent system. The purpose of the statute was to suppress patent letters establishing monopolies, except those granted to "the true and sole inventor of any manner of invention unknown in the realm". But even when England became the most advanced country and the cradle of the Industrial Revolution, the grant of patents remained a prerogative of the Crown, and by no means a right of the inventor.

The laws, adopted at about the same date (1790 and 1791) by the United States and France, opened a new era of the patent legislation. In the particularly explicit wording of the French law, every inventor is owner of his invention, and the purpose of the law is only to lay down the conditions of this right. This conception became the basis of the different patent laws in force nowadays in practically all countries.

The objective conditions that an invention must fulfil in order to be eligible for patenting are also nearly world-wide:

- the invention must have a technical objective
- it must be new, that is to say, not contained in the 'state of the art' at the date of filing the application
- it should not derive obviously, for a person skilled in the art, from the same 'state of the art'.

One of the most striking features of the patent system for the last century has been the development of internationalisation. The first decisive step was taken in 1883, and has been completed by the European Patent Convention and the Patent Co-operation Treaty (both operative 1978).

Why patented inventions are fully described in a document open to the public

The requirement of a full description of the invention is a consequence of the fact that it has, from the very beginning, been necessary to check the claims of the applicant in order to avoid granting exclusive rights to those who do not merit them. But it implied by no means that this description should be made public with more detail than was necessary for warning the public of the existence of a certain privilege. But with the ideas prevailing in the last years of the eighteenth century, new motives appeared for a full description of the invention to the public:

- new inventions, being a valuable contribution to technical knowledge, should be publicly disclosed, in order to enhance progress
- the inventor seeking the advantages of the patent law should, as a counterpart, give a complete account of the invention, omitting no means of realisation (this relation between full description and exclusive rights has often been described as a contract between inventor and State)
- the competitors of the owner of the patent should know exactly where they stand, whether they run the risk of infringing the patent, or are free to use the techniques neighbouring the patented one.

It is obvious that if the disclosure of the invention is beneficial for competitors and for the public, it becomes an obligation for the inventor. Of course, the inventor is not compelled to apply for a patent. He remains free to choose secrecy and try to keep the invention secret throughout the whole process of exploitation of the invention, but this is impossible for certain kinds of inventions. It is easy to predict that inventors would like to take advantage both of the patent and of secrecy; this is often expressed candidly by unexperienced inventors, but it is also attempted by clever specialists. Patent laws are accordingly particularly severe against dissimulation of the means actually used for making the invention.

Problems related to the number of patent documents

Towards the end of the eighteenth century, the number of patents granted in the world amounted to a few score per year. In 1991, the number of patent documents that represent the state of the art reaches more than thirty million, with an annual increase of eight hundred thousand. Even if one takes into account the fact that the same invention gives birth to several patent documents, due largely to the fact that patents are as a rule limited to the territory of the state that granted the patents, the number of patent documents remains huge and impossible to manage without special classification and diffusion methods.

Date of publication of patent documents

The exclusive right granted to the inventor by the patent are protected as from the date of the first filing of a patent describing the invention. This is the earliest date conceivable for publication, since divulgation before first filing date may prevent valid patenting by destroying novelty.

But such an early divulgation is apt to be detrimental to the inventor, especially if he wants to withdraw his first application in order to, for instance, file a more complete and better drafted one. For this reason, publication immediatly following first application is not compulsory, and may be made only by the inventor himself. Many patent laws provide for publication of the patent application eighteen months after filing date. This delay has been deemed long enough to preserve the inventor's rights of withdrawing or amending his application, and short enough for third parties and the general public. Still, some patent laws, especially the U.S. patent law, do not provide for publication before grant of the patent.

Patenting in chemistry, pharmacology and biotechnology

At the time of the first patents, most inventions were in the field of machines or mechanical processes, but no specific exclusion was therefore prescribed against patenting certain kinds of inventions. But as a result of scientific achievement in the fields of chemistry, pharmacology and biotechnology, it appeared that patents in these areas raised a number of specific difficulties.

Technical difficulties

Some of the rules of patenting are difficult, or even impossible to apply to certain inventions. For instance, living beings used or produced by a process are not always indentified by a taxonomic description. This difficulty has been avoided by the practice of deposit in public collections of living organisms. The particularities of plant varieties, especially when obtained by selection, hybridation, etc., have led to special laws, differing more or less from patent laws.

Economic difficulties

Patenting a chemical product *per se* has been looked upon as giving to the inventor an excessive economic advantage, since such patents covering the product itself, whatever the processes of production may be, can be of very wide economic impact. For a long time, many countries prohibited patenting chemical products *per se* and allowed only patenting of chemical processes.

Ethical difficulties

The above-mentioned economic problem combined with the purpose of pharmaceutical products and biological products, raised ethical objections of several kinds:

- at a time when 'patented drugs' were of a rather dubious value, the French legislature (1844) placed pharmacy out of the field of patenting, in order to avoid the public being induced to buy those drugs because of their patents, that is seeming to have received official approval.
- More recently, and with the progress of pharmacology, it has been deemed damaging to public health to grant patents for pharmaceuticals, with the idea that such patents might prevent the diffusion of new remedies at reasonable prices.

Other ethical objections have been raised against patenting living beings, but they seem to rely mainly on a misconception of the true nature of patents. On the whole, the general tendency has been toward removing these different kinds of difficulties and granting patents for all these types of inventions. But of course, the relatively late dates of patenting of these inventions has the consequence that patent documents in these areas do not go back as far in the past as in other fields of technology.

Patent databases and other media

A simple overview looks like *L'inventaire de Prévert* and reaches 56 databases including patents, and more than 20 on CD-ROM. But what is better is to refer to WIPO's document PCPI/EXEC/VIII/10 July 5, 1991 (Permanent Commitee on Industrial Property Information, Working Group on General Information). The annex 4 of this document 'Survey of the available databases, their services and access conditions' gives a complete answer and, more than an overview of the patent databases available, their services and access conditions: 35 online patent databases are available on about 10 commercial vendors.

Why so many databases?

Threefold significance of patents

A patent is a document granted by a state agency (or by a regional agency acting for several states) describing an invention and creating a legal exclusive right for the patentee. According to this right, the subject matter of the patent (product or process) cannot be produced, used, sold or imported without the agreement of the patentee.

It appears clear from this definition that a patent relies basically on technology, since creations of a non-technical character cannot be patented.

The exclusive right of the patentee calls for a complex practice of legal procedural rules governing the grant of a patent and the possible conflicts between the patentee and third parties. The legal character of patents gave birth to a whole profession, organised and regulated (patent agents).

But finally, the purpose of the patent system is to introduce a temporary distortion of the laws of free competition in order to enhance technical progress through the hope of eventual profits.

The economic ends of patents are at two levels:

- on a practical ground, the fact that a patented product has unique characters, protected by laws against imitations, provides a telling argument for commercial and marketing agents; they are also apt to suggest technical improvement to the products of their organisations when they have to compete with patented products of other organisations.

- at a strategic level, the general policy of public and private enterprise is concerned with the powerful means of securing a dominant position that is obtained through research and patenting; economics is the root and the aim of the patent system, whose results are of importance for both private enterprise and states.

The three-fold character of the patent system implies that patent documents are of interest to different groups of people.

Technicians find in patent documents the most detailed, complete and earliest account of inventions. On a general level, knowledge of patent documents ought to be a part of the culture of every engineer, technician and, in many cases, scientist. From a more practical point of view, every researcher is able, with patent documents, to learn about what others have acheived or have attempted in the fields of his own research. Thus, useless research leading to something that has already been invented may be avoided. The previous work on a problem proves stimulating for devising new — and patentable — solutions to this problem.

Commercial and marketing agents find in patent documentation information about what their competitors are going to put on the market. They will also be made aware of the extension of protection of the patents of their own organisation and of those of their competitors. Since an invention is practically never protected in all countries, the knowledge of the range of protection (both geographical and in time) is an essential element of commercial strategy. The knowledge of the patents of a possible partner is necesary, when any form of association, merging or simply licensing is contemplated.

Lawyers are responsible, during the instruction of a patent application by the patent office, for complying with the intricate provision of patent laws and implementing regulations. They have to answer to objections of a legal character formulated by the patent office's examiners. Furthermore, they have to delimit the scope of protection of patents, which is of the highest importance in the case of a suit for infringement. They are, of course, in charge of drafting legal documents and contracts concerning the ownership and exploitation of patents.

Economists. Patent document statistics, especially when broken down according to country of origin and technical groups, are an indicator of the results of R&D in different countries and in different industrial branches and even enterprises. For the managers of industrial firms, patents are not only a source of eventual profits, which has to be calculated taking into account the cost of obtaining, maintaining and defending patent rights. But they are also an often decisive element in competition. By securing an exclusive right to a process or product, an organisation is able to gain parts of a market and to negotiate favourable agreements with other organisations. Even if some patents have an enormous market value, and many none, a long-term policy of research and patenting is not gambling, because many minor improvements to a technique are a surer way to success than haphazard 'flashes of genius'.

Conclusion

Such is the instructional wealth of patent information, so wide are the range of needs it provides with an answer, that one ought to be thankful for the fact of its cross-checking through multifarious producers and vendors. The arrangement of data in each database, the search language, technical fields, selected mother-tongues and classifications, span of time covered, all justify what, at first sight, might appear to be a redundancy, but is, in fact, an invaluable richness.

Author Index